Movement Science

Foundations for Physical Therapy in Rehabilitation

Second Edition

Janet Carr, EdD, Dip Phty, FACP
Associate Professor
School of Physiotherapy
The University of Sydney
Lidcombe, Australia

Roberta Shepherd, EdD, Dip Phty, FACP
Professor
School of Physiotherapy
The University of Sydney
Lidcombe, Australia

AN ASPEN PUBLICATION®
Aspen Publishers, Inc.
Gaithersburg, Maryland
2000

Bay State

The author has made every effort to ensure the accuracy of the information herein. However, appropriate information sources should be consulted, especially for new or unfamiliar procedures. It is the responsibility of every practitioner to evaluate the appropriateness of a particular opinion in the context of actual clinical situations and with due considerations to new developments. The author, editors, and the publisher cannot be held responsible for any typographical or other errors found in this book.

Library of Congress Cataloging-in-Publication Data

Movement science: foundations for physical therapy in rehabilitation / [edited by] Janet Carr, Roberta Shepherd.—2nd ed.
p. ; cm.
Includes bibliographical references and index.
ISBN 0-8342-1747-3
1. Physical therapy. 2. Neuromuscular diseases—Patients—Rehabilitation.
I. Carr, Janet H. II. Shepherd, Roberta B.
[DNLM: 1. Movement Disorders—rehabilitation. 2. Models, Neurological. 3. Nervous System Diseases—rehabilitation. 4. Physical Therapy. WL 390 M9358 2000]
RM700 .M68 2000
615.8'2—dc21
99-089080

Orders: (800) 638-8437
Customer Service: (800) 234-1660

Editorial Services: Joan Sesma
Library of Congress Catalog Card Number: 99-089080
ISBN: 0-8342-1747-3

Printed in the United States of America

1 2 3 4 5

2/26/01

Table of Contents

Contributors

A.M. Gentile, PhD
Professor of Psychology and Education
Department of Biobehavioral Studies
Teachers College, Columbia University
New York, New York

James Gordon, EdD, PT
Professor of Practice
Program Director
Program in Physical Therapy
New York Medical College
Valhalla, New York

Jean M. Held, EdD, PT
Associate Professor
Chair
Department of Physical Therapy
University of Vermont
Burlington, Vermont

Preface to the Second Edition

The first edition of this book was an early attempt to argue for the relevance to the clinic of scientific findings related to human movement. The new edition continues this theme and the material has been updated in light of new experimental findings in science and clinical practice.

The second edition closely follows the format and approach used in the first edition. An epilogue has been appended to Chapter 1 since Gordon's historical perspective on the assumptions underlying the dominant (in 1987) methodological and theoretical facilitation approach is still relevant today. Encouraging signs in neurorehabilitation indicate a major paradigm shift away from the old facilitation model toward a more dynamic and interactive training model. The process of scientific change from an old to a new model, as Gordon[1] points out, becomes necessary when old assumptions are no longer able to solve the present problems. However, during the transition from old to new, there will always be attempts to reconcile and hybridize the two competing models, although this is rarely an effective solution.[2,3]

Chapter 2 has been revised extensively to incorporate additional information on the prevention of cardiovascular and neuromotor disuse changes that are secondary to immobility, on strengthening exercises, task-specific training, and the need to increase the time individuals spend in physical activity. Developments in scientific and clinical research since the first edition presented a challenge in updating the content of this chapter without expanding it to unrealistic proportion. In reviewing the literature, a sense of optimism arises from evidence presented in many recent studies. These studies demonstrate the effects of task-specific exercise and training and the potential for improved outcomes with more modern methodologies. Research studies included in this chapter serve two purposes: to present an example of research done into an issue to substantiate a point; and to provide research references for the clinician, student, and teacher who want more information about a topic.

Chapters 3 and 4 have been revised and updated to include new research on skill acquisition in healthy subjects and factors influencing recovery of function after brain damage. The material in Chapter 3 is particularly relevant to an exercise and training model of rehabilitation that is emerging around the world in which the patient is an active participant. Chapter 4 provides insight into the potential for the process of neurorehabilitation to impact directly and indirectly on neural reorganization.

What place will physical therapy hold in neurorehabilitation in the future? Methods of delivery are being challenged by funding bodies who are less prepared to fund lengthy periods of rehabilitation unsubstantiated by evidence of effectiveness. It is evident to us that there is an urgency for a change to a more scientifically rational, demonstrably effective, and more economically viable delivery of rehabilitation. Therapists can no longer afford to deliver their "personal" rehabilitation but rather need to follow guidelines and protocols defined in reference to scientific findings and evidence of effectiveness. There is sufficient evidence of what is effective and what is not to make considerable change both to methodology and delivery. As Rothstein[4] has urged, once some evidence indicates what works and what does not, it is critical that such evidence is used in clinical practice.

REFERENCES

1. Gordon J. Assumptions underlying physical therapy intervention: theoretical and historical perspectives. In: Carr JH, Shepherd RB, eds. *Movement Science: Foundations for Physical Therapy in Rehabilitation.* Rockville, MD: Aspen Publishers; 1987.
2. Abernethy B, Sparrow WA. The rise and fall of dominant paradigms in motor behaviour research. In: Summers JJ, ed. *Approaches to the Study of Motor Control and Learning.* Amsterdam, The Netherlands: Elsevier Science; 1992.
3. Kuhn TS. *The Structure of Scientific Revolution.* Chicago: University of Chicago Press; 1970.
4. Rothstein JM. It is our choice! *Phys Ther.* 1997;77:800–801.

Janet H. Carr
Roberta B. Shepherd

Acknowledgments

Janet Carr and Roberta Shepherd wish to thank the people who so kindly agreed to be photographed for this book; the physiotherapy department at Prince Henry Hospital, Sydney; Colleen Canning for her insightful comments on an early draft; and David Robinson, who took the photographs.

The authors and publishers also wish to express their appreciation for being granted permission to reproduce figures, tables, and/or text as indicated throughout the book.

Introduction to the First Edition

Physical therapy is in part an applied science, and the physical therapist with the relevant postgraduate qualifications and working with motor-disabled people could be considered an applied movement scientist. In this role, the physical therapist has a unique opportunity to improve rehabilitative training programs and to collect data on the effects of disease and trauma on human motor behavior. As an *applied movement scientist*, the physical therapist extrapolates information from movement science, deriving inferences from this information and devising strategies for the clinic. In addition, as *movement scientist*, the physical therapist will also be involved in research into human movement.

Janet Carr and Roberta Shepherd have been developing a new therapeutic model for rehabilitation out of movement science theory and research. In a book published originally in 1982, they illustrated how this information may be applied to the rehabilitation of people whose movement dysfunction is the result of stroke. It has been customary in physical therapy for approaches to therapy to be developed by clinicians in an *inductive* manner—clinical findings of interest leading to a search for a theoretical explanation. This was largely due to the relative lack of a scientific body of knowledge related to movement. The increasing volume of movement-related research over the last decade has, however, made possible the development of therapeutic models by a *deductive* process—clinical models being developed out of a particular theoretical base. The therapeutic model described by Carr and Shepherd is being developed by this latter process.

The first chapter of this book is a theoretical and historical perspective of physical therapy in neurological rehabilitation, in which the relationship between, on the one hand changes in scientific attitude and new scientific knowledge, and on the other the development of new therapeutic models, is discussed. Gordon emphasizes the need to question the implicit and explicit assumptions underlying both clinical theorizing and practice in the light of increasing scientific knowledge about motor control processes, the effects of lesions, and recovery processes. Although neurological rehabilitation is used as an illustration, the therapeutic approaches he describes are widely utilized in general rehabilitation and hence his perspective of critically analyzing the assumptions underlying physical therapy practice is applicable in a more general sense.

In Chapter 2 Carr and Shepherd set out to demonstrate how to use theoretical material from movement science to develop a therapeutic model for rehabilitation. In this model, a patient's motor performance on attempting a particular task (e.g., standing up) is compared to normal performance, information about which comes from biomechanical and other research evidence. Clinical intervention is largely a training process in which the physical therapist assists the patient to activate those muscles which are difficult to contract, encourages practice of simple exercises designed to improve control over those muscles, and teaches the patient how to perform relevant motor tasks concentrating on the performance of those components (or biomechanical parameters) which are both critical to that task and difficult to perform. Emphasis is thus on the patient as active learner and problem solver. This model is illustrated in this chapter with relation to stroke rehabilitation but is intended to demonstrate a wider application of such material.

Clinical examples are included throughout the second chapter in order to make clear the process of deriving implications for the clinic from scientific material (the deductive process). The role of the physical therapist working as applied movement specialist will increasingly depend on reading scientific material, seeing what is relevant to clinical problems and the training process, devising therapeutic strategies from this material, and testing whether they lead to improved performance in the functional tasks required. This process must be seen to be in a continual state of development since the basic sciences that form the theoretical framework for rehabilitation are themselves in a constant state of development.

One assumption in the second chapter is that current knowledge of how people learn (or acquire skill in) motor tasks is relevant to the motor-disabled person who must relearn what were once habitual everyday tasks. Hence in the third chapter, Gentile presents one perspective from which the process of learning can be viewed. This chapter provides insights relevant to the rehabilitation of all movement disabled people, but in terms of this book, provides more detailed information about those aspects of motor behavior, neuromotor processes, and task learning which are of particular relevance to a model of rehabilitation in which the patient is an active learner.

Another assumption in the second chapter is that the system is capable of a degree of recovery and that events subsequent to a lesion may affect recovery processes. Hence, in the fourth chapter Held reviews recent experimental findings related to the brain's adaptive capacity in order to illustrate the relevance to clinical attitudes and strategies of understanding recovery processes. In particular, she points out the evidence supporting the need for the rehabilitation environment to be stimulating, for therapists to have positive attitudes toward recovery, and for early training to be task-specific.

The reader may like to select his or her own path through this book. However, the chapters have been ordered in such a way as to take the reader on a trip that starts by discussing how clinical practice is based on certain theoretical assumptions and how therapeutic models are developed and goes on to illustrate the development of a new deductive model and its application to a particular set of problems. The last half of this intellectual journey shows the reader how, by observing and describing motor behavior both in skill acquisition and following lesions of the CNS, the teacher of motor behavior can learn how to structure practice and to organize the environment to facilitate learning and recovery with the emphasis on the active participation of the learner (or patient). Since each chapter reflects the author(s) own way of using his or her knowledge to support the general concept, each chapter can also be read as a distinct entity. However, for those who take the trip as ordered, the editors hope that a coherent theme will be evident throughout the entire work.

The material presented in this book is considered by the authors to be relevant to physical therapy students both at the undergraduate and graduate level. The graduate level student in particular will want to read about the theoretical framework in more depth; hence each chapter's reference lists will provide a suitable basic reading course in the area of movement science. In planning this volume we hoped that our combined interests in movement science and our own individual areas of knowledge would produce a book of interest to all those working in the rehabilitation field and perhaps provide the rationale for the development of new therapeutic attitudes and strategies.

Janet H. Carr
Roberta B. Shepherd

Chapter 1

Assumptions Underlying Physical Therapy Intervention: Theoretical and Historical Perspectives

James Gordon

How do new treatment approaches develop? To attempt to answer this question, the development of neurophysiological approaches in physical therapy will be considered from a historical perspective. The overall purpose of this chapter is to develop a framework for evaluating both established approaches and new ideas concerning how patients with neurological problems should be rehabilitated.

Any therapeutic approach aimed at rehabilitation of neurological patients is based on *assumptions* regarding how the central nervous system (CNS) controls movement. These assumptions are often *explicitly* stated by the originators or proponents of a particular approach. For example, most of the established neurophysiological approaches to treatment assume that recovery from brain damage follows a predictable sequence, similar to the development of normal motor behavior in infancy. This assumption leads to specific recommendations concerning the progression of patients through a sequence of treatment activities. Most of the assumptions underlying therapeutic approaches are, however, *implicit*, that is, taken for granted. For example, it is often taken for granted that practice of some act will lead to greater skill. The underlying assumption is that repetition in some way "strengthens" appropriate synaptic connections in the CNS. Everything that we as therapists do in the clinic is based on assumptions such as these, although we may not think about them as we treat patients and we would most likely have difficulty in identifying them if asked. In order to develop a meaningful critique of our therapeutic approaches, it is first necessary to identify the assumptions on which they are based. From these assumptions, we can begin to understand the overall *theoretical model*, that is, the set of assumptions, both explicit and implicit, guiding a particular therapeutic approach. A framework for organizing and comparing the assumptions underlying particular therapeutic approaches is suggested later in this chapter.

Why do we need to change our therapeutic approaches? An answer to this question can take two forms: (1) because current approaches are not adequate to solve clinical problems as they are perceived by therapists and (2) because the theoretical assumptions underlying current approaches do not fit with current knowledge. It is when current approaches are perceived as inadequate both practically and theoretically that optimal conditions exist for therapeutic innovation.

Where do new ideas for therapeutic intervention come from? Ideas come both from theoretical advances (i.e., in the neurosciences and behavioral sciences) and from practical experimentation (i.e., in the clinic). However, specific ideas for therapy must come from therapists, those people directly involved in attempting to find solutions for patients' problems. Therapists often look to scientists for answers to specific questions about what to do or for direct ideas concerning how to translate experimental results into therapeutic techniques. This is impractical and leads to unrealistic expectations concerning the relationship between scientists and clinicians. The important role of scientific theory is to provide an underlying model that guides the integration of practical ideas into a coherent treatment philosophy. This model provides general ideas or guidelines concerning how the CNS controls movement, what goes wrong following brain injury, how recovery from brain injury occurs, and what mechanisms can be used to enhance recovery. Thus, we cannot look to the basic sciences for the specific new ideas about how to treat patients. Science provides us with general models; *we* must work out their practical implications.

HISTORICAL DEVELOPMENT OF NEUROPHYSIOLOGICAL APPROACHES

New ideas do not come about by spontaneous generation. More often than not they are reactions to older ideas, and they arise because the older ideas are perceived as no longer able to solve present problems. As an example, it is useful to consider how current established therapeutic approaches to the treatment of neurologically impaired patients were developed. By established approaches, I mean those based on the work of the Bobaths, Brunnstrom, Rood, Fay, Kabat, Knott, Voss, and many others.[1–6] These approaches are often referred to as "neurophysiological" approaches to treatment;[7] for this reason, I will refer to them as "neurotherapeutic" approaches.

What were the dominant treatment philosophies before the development of the neurotherapeutic approaches? To answer this, we must place ourselves in the historical situation faced by the developers of these approaches. Before the 1950s, when the neurotherapeutic approaches were in their infancy, the practice of physical therapy was to a great extent concerned with the treatment of patients with poliomyelitis. Indeed, the growth of physical therapy as a profession in the first half of this century was in part due to the need for rehabilitation of the thousands of children and adults stricken with this disease.[8–10] The dominant treatment approach, the "state of the art," was referred to as muscle reeducation.[11] As its name implies, this was a molecular approach, focusing on individual muscles. Manual muscle testing was used to identify weakened muscles, and patients were taught specific exercises in order to strengthen these muscles. In this technique, therapists focused the conscious awareness of the patient on an isolated muscle. Here, for example, is an excerpt from a description of the Kenny method of muscle reeducation:[9]

> The conscious mind must be made aware of the muscles and must visualize contraction of the non-functioning muscle when the joint is moved in order that orderly and coordinate joint action will ensue. To accomplish this result, the patient must be accurately instructed in the location of muscles and particularly in the insertions of muscles. He must understand the resultant action of the joint and parts when the insertion of the muscle is pulled upon by contraction of the muscle.

Muscle reeducation was an effective technique for treating poliomyelitis patients, who often had focal paralysis of muscles scattered throughout the body. At the same time, the rehabilitation of neurological disorders such as hemiplegia and cerebral palsy was considered largely an orthopedic problem, with surgery, bracing, and "reeducation" the dominant treatments.[12–14]

What happened, of course, is that the Salk vaccine began to be used in the mid 1950s, and, within a few years, poliomyelitis was largely eliminated as a serious problem. As therapists recruited in the battles against poliomyelitis turned their attention to other patients, especially those with neurological disorders, it became obvious to many that muscle reeducation and standard exercise therapies were inadequate. Poliomyelitis, while technically a CNS disorder, is properly a disease of the "lower motor neuron," equivalent to a lesion of the peripheral nerve. Most CNS disorders, including those resulting from strokes and cerebral palsy, damage higher centers of the nervous system where a more complex integration of the neural signals responsible for control of movement takes place. Thus, although poliomyelitis may be amenable to a molecular approach such as muscle reeducation, most other neurological disorders are not. The impairment is simply not expressed in individual muscles but rather in the construction of whole purposeful movements.

Here a remarkable process began. Therapists actually began to study how the CNS works and to read and discuss contemporary thinking in neurophysiology, with the express purpose of generating ideas about how to better treat patients with brain damage. The result of this process—the development of several related neurotherapeutic approaches—brought about a true revolution in the way neurological patients were rehabilitated. Where previously they had been considered an orthopedic problem, to be braced or operated on, or strengthened where appropriate, true remediation of these movement disorders was now considered possible. The key to this revolution was the change from viewing their deficit as peripheral, with the concomitant emphasis on specific treatment of individual joints or muscles, to a view of their deficit as central and neural in origin. The crucial idea was that the CNS itself should be the focus of treatment, not the peripheral effects of the CNS lesion. Because the deficit was in the CNS, the neurotherapeutic approaches aimed their treatment at the CNS itself.

The process of working out the neurophysiological approaches to treatment was both exciting and confusing. Several approaches developed, and proponents of each argued their respective merits in meetings and journals, and in the clinics themselves. This period of ferment in physical therapy continued through the 1950s and much of the 1960s and culminated in the extraordinary set of sessions at Northwestern University in 1964—the Northwestern University Special Therapeutic Exercise Project (NUSTEP).[6] Since then, the ferment has died down, the proponents of the various approaches have made peace with each other, and the emphasis has shifted to refining the techniques developed by the Bobaths, Rood, and others. After 30 years, the revolutionary ideas of the neurotherapists have become the *status quo*, the established approaches to the treatment of neurologically impaired patients.

COMMON ASSUMPTIONS UNDERLYING THE NEUROTHERAPEUTIC APPROACHES

Although there are certainly substantive differences among the various neurotherapeutic approaches, there are strong similarities in the theoretical assumptions on which they are based. Here I will discuss what I believe are the most important of the common assumptions. I will refer to the set of assumptions underlying these approaches as the *facilitation model*.

The brain controls movements not muscles. This aphorism, adapted from John Hughlings Jackson, the famous 19th century neurologist, expresses a concept central to the neurotherapeutic approaches. It is a specific repudiation of the muscle reeducation model and implies that there should be a more holistic emphasis on movement patterns, rather than a molecular emphasis on manual muscle testing and the strengthening of individual muscles. Although it is clearly an oversimplification of Jackson's views[15] and is problematic if taken literally,[16] it properly emphasized that the CNS acts to link together contractions of different muscles in order to produce skilled movements. All of the neurotherapies assume that lesions of particular areas of the CNS will lead to disordered patterns of movement, rather than simply paralysis or weakness of individual muscles. Thus, while treatment philosophies may differ, all approaches focus on movement patterns. Some advocate attempting to suppress "pathological" movement patterns;[17] others suggest making use of pathological patterns as a first step toward regaining normal patterns.[18] The common assumption is that an abnormal pattern of movement results from the lesion itself, rather than from the patient's attempt to compensate for the lesion. The implications of this assumption will be discussed later.

We can alter, or *facilitate*, a patient's movement patterns by applying specific patterns of sensory stimulation, especially through proprioceptive afferent pathways. This assumption had its roots in the dominant stimulus–response methodology of behavioral psychology as well as the reflexology of contemporary neurophysiology (exemplified by the work of Sherrington and Magnus in the early part of the century). It was also based on practical experience: therapists recognized that it was possible to elicit specific movement patterns in patients by the careful application of appropriate sensory stimuli. It is easy to see why sensory stimulation as a therapeutic technique was so seductive to the therapists developing neurophysiological approaches. It offered access to the CNS—a direct way of getting to the source of the problem. There was a real hope in the early days of the neurotherapies that facilitation would be a powerful tool, and much effort was devoted to working out the effects of different types and sites of stimulation. The implicit assumption was that sensory stimulation would produce lasting effects, that is, it would facilitate changes in the CNS. The mechanisms underlying such lasting effects have never been clearly defined, and, as will be discussed later, many therapists now doubt whether sensory stimulation by itself can bring about long-lasting changes at all. Nevertheless, sensory facilitation of normal movement patterns is a cornerstone of the neurotherapeutic techniques.

The CNS is hierarchically organized, with higher centers normally in command of lower centers, which in turn control primitive and more automatic behaviors. This assumption is again borrowed from Jackson and is a central idea in 20th century neurophysiology.[19] The importance of this idea is in its implications for dyscontrol,[20] that is, the nature of the deficit produced by damage to the CNS. The implication of this assumption is that damage to higher centers, especially the cerebral cortex, leads to dissolution, or *release* of lower centers from higher control. The result is that the patient functions at a more primitive level and is unable to suppress the automatic movement patterns controlled by these lower centers. Thus, abnormal patterns of movement (as well as disorders of tone, such as spasticity) are seen as resulting from a lack of inhibitory control by higher centers. Patients with damage to higher centers are unable to produce "fractionated" movements of the extremities; they are locked into certain stereotyped movement patterns. The release concept of dyscontrol implies that treatment should be aimed at reestablishing inhibitory control by higher centers.

Recovery from brain damage follows a predictable sequence that mimics the normal development of movement during infancy. This follows from the previous assumption (hierarchical organization). Both development and recovery are viewed as progressive encephalization, with higher centers gaining control of more primitive lower centers. This assumption has had perhaps the most important effect on the way we structure our treatment approaches to neurological problems. It can be found as an axiom in almost every textbook on therapeutic exercise. It has led to a direct and sometimes rigid application of a developmental sequence to the progression of therapeutic exercises.

Primacy of the neurophysiological explanation. The overriding assumption guiding the development of the neurotherapies, which is almost always taken for granted, is that all motor phenomena associated with brain damage have a neurophysiological basis. As mentioned previously, this assumption was a reaction to the older approaches to treatment that focused on the peripheral expression of brain damage in the musculoskeletal system rather than on the lesion to the brain itself. Thus, for example, rather than treating the effects of spasticity by bracing or surgery, it was believed that the correct approach was to attempt to modify the abnormal tone at the source by application of specific sensory stimulation. Similarly, if a treatment such as manual stabilization of proximal joints helps a patient to achieve better control of distal movements, the explanation for this phenomenon is viewed as neurophysiological: normal patterns of neural functioning have been facilitated by the therapist. This assumption, the primacy of the neurophysiological explanation, undoubtedly had a positive influence on the training of physical therapists, leading to a major emphasis on the neurosciences within physical therapy curricula. On the other hand, it may have had the unfortunate effect of promoting a one-dimensional view of brain-injured patients, resulting in a lack of concern for other useful perspectives, including biomechanics, muscle biology, and the behavioral sciences.[21]

RELATIONSHIP BETWEEN SCIENTIFIC THEORY AND CLINICAL PRACTICE

What is the role of basic science in determining the way we develop and change treatment strategies? The somewhat oversimplified historical summary presented here provides an example of how science and clinical practice interact. What are the lessons to be learned from this historical perspective?

Science provides a guiding theoretical model. Basic science rarely, if ever, gives us specific techniques. Rather, it provides the guiding set of assumptions about how the CNS controls movement, what goes wrong when the CNS is damaged, and the possible mechanisms of recovery from brain damage. The founders of the current established neurotherapies made observations of patients and the effects of certain techniques. But these observations were not simply random. They were systematic, that is, based on an overall framework or model. Therapists categorized those observations and generated new ideas based on the set of assumptions they had borrowed from the basic sciences. In this regard, it is relevant that more than half of NUSTEP consisted of basic science presentations.[6] For instance, much attention was given to the proprioceptive system, especially muscle spindles, and its role in movement. The strong emphasis within neurophysiology on the role of this system in motor reflexes translated to an assumption that stimulation of the proprioceptive system would be a useful tool in rehabilitation.

Because we are removed in time from the original development of the neurotherapies, it is easy to take the assumptions on which they are based for granted. However, for those therapists who developed the neurotherapies, contemporary neurophysiological theory played a central role in the formulation of the overall model guiding their observations and practical experimentation. During the 1940s, 1950s, and 1960s, a fundamental shift in the thinking of physical therapists had taken place, from an emphasis on individual muscles and joints to a focus on internal neural processes. The reason for this shift was that previous models, while useful for the treatment of disorders such as poliomyelitis, were seen as not able to solve the problems of therapists attempting to treat patients with central neurological deficits.

Practical needs determine the validity of the theoretical model. A theoretical model is not simply right or wrong. It is valid only insofar as it is useful. The founders of the neurotherapies did not adopt this neurophysiological model (what I have referred to as the facilitation model) because it was correct in some absolute way but rather because it led to a whole universe of ideas about how to treat neurological patients in new and effective ways. One might say that the facilitation model was more correct than the muscle reeducation model because it was better suited to solving the problems of therapists as they perceived them at the time.

Here it may be useful to note the similarity of the process of developing the neurotherapies to the description by Thomas Kuhn of what he referred to as scientific revolutions.[22] Kuhn proposed that a particular scientific discipline normally operates within a dominant methodological and theoretical approach or set of assumptions, which he referred to as the paradigm. Normal science explores the implications of the paradigm and works out the problems that the paradigm is particularly well suited to solve. There comes a time, however, when the paradigm is no longer able to solve the problems that are perceived as important by the members of the discipline. For example, medieval astronomy was dominated by the belief that all planets, including the sun, revolved around the Earth. However, as astronomical methods improved, this belief became increasingly unable to explain the observations of astronomers; this led to a period of crisis, with competition between new ideas. Ultimately, a new set of assumptions won out that was better able to explain phenomena that could not be dealt with under the old paradigm. In the case of astronomy, the new paradigm, developed by scientists such as Copernicus and Galileo, was based on the revolutionary assumption that the planets revolved around the sun.

Thus, Kuhn was proposing that scientific revolutions are brought about when scientists undergo a radical change in their overall set of assumptions, what he referred to as a paradigm shift. This does not necessarily come about gradually but rather in a period of crisis, when it is perceived that the old assumptions are no longer useful. What brings about a paradigm shift is thus a change in scientists' perception of what problems are important to solve. Moreover, the new paradigm that wins out does not do so simply because it is correct but because it is the one best able to solve the problems perceived as most important. Although Kuhn's theory of scientific revolutions was intended as a way of understanding how the basic sciences develop, it also provides a useful framework for understanding how innovation occurs in clinical sciences, such as physical therapy.

CURRENT FACTORS PRODUCING A DYNAMIC FOR CHANGE

Although recognizing that it is difficult to analyze the current situation within a historical perspective, I think it is fair to state that there is a crisis of sorts in the approach of physical therapy to the treatment of neurological patients. Established approaches are no longer viewed with the same optimism as they once were. An indication of the disillusionment with neurotherapeutic approaches is the popularity of orthopedic therapies, especially among younger therapists. When once the most dynamic area of physical therapy was rehabilitation of the neurologically impaired, courses in manual and orthopedic therapy now draw considerably larger audiences than courses in the neurotherapies. It is worth considering, therefore, the current factors that might lead to innovation in treatment approaches for the neurologically impaired.

Changing Roles of the Rehabilitation Therapist

Although the increased interest in orthopedic therapies would seem to indicate that there is less need for neurological rehabilitation, this is hardly the case. In modern society, tertiary care has assumed increasing importance. Today more patients are surviving head injuries, brain tumors, strokes, birth injuries, and premature deliveries. Furthermore, there is increasing emphasis on improving the quality of life of the elderly population, who are most likely to suffer strokes and other neurological problems. Thus, if anything, there has been an intensification of the need for effective treatment approaches in neurological rehabilitation. With this increased need has come requirements for increased accountability, especially from government agencies and third party payers. Therefore, at the same time that the need for neurological rehabilitation is intensifying, we are being asked to justify our treatment strategies, both in terms of their scientific basis and their practical effectiveness.

Dissatisfaction with the Facilitation Approach: The Problem of Functional Carry-Over

As we all discover early in our careers, it is easy enough to "facilitate" a certain pattern of movement. What is difficult is to get patients to use that pattern when they are actually carrying out some functional activity. This is the fundamental challenge facing rehabilitation therapists. There is constant tension in therapy between the desire to achieve true restitution of normal movement patterns and the practical need to train patients to carry out essential functions in their daily lives. Unfortunately, facilitation approaches have not directly confronted the problem of how to achieve functional carry-over. Too often there is simply the assumption that the treatments we perform will somehow mysteriously translate into functional improvement for the patient. Therapeutic exercise and training in activities of daily living are segregated and performed at different times and often by different therapists. The apparent contradiction between facilitation of normal movement and independent function in daily activities has been a major factor in producing disillusionment among therapists, especially as we are increasingly faced with the requirement to discharge patients earlier. Essentially, the facilitation approaches promised more than they could deliver. The hope was that we could reinstate normal movement patterns. The reality is that, even when we succeed in accomplishing this, we find that patients use movements different from the ones we teach them when they are confronted with functional tasks in meaningful environments. New approaches must deal directly with this problem. Interestingly, the neurophysiological approaches were an advance over the muscle reeducation approaches in that they stressed total patterns of movement, a higher functional level than the isolated muscle contractions stressed by muscle reeducation approaches. The next advance must be to combine training of normal movement patterns with emphasis on their use during normal functional activities.

Theoretical Advances: The Motor Control Perspective

In the past 20 years there have been great advances in our understanding of the neural bases of movement. There has been a shift in emphasis from purely anatomical and physiological explanations of the motor system to what Greene has referred to as a *task-oriented approach*.[23] This has come about because of the recognition that it is not enough to develop a wiring diagram of the CNS (even if this were possible) in order to understand how it works. Furthermore, much physiological experimentation into the workings of individual elements of the CNS has been frustrated by the lack of basic agreement on the functions of these parts. A good example of this is the muscle spindle. After at least 40 years of intensive study of the structure and function of this receptor, there is still no consensus on its function in normal movement control.[24] A task-oriented approach assumes that before we can understand how the CNS controls movement we must first understand the problems faced by the CNS in controlling movement. What problems does the CNS have to solve in order to perform specific motor tasks successfully?

An example of a problem faced by the CNS in controlling movement was first formulated by Bernstein,[25] who asserted that the principal problem faced by the CNS was the enormous number of joints and muscles in the human body. Because of the large number of degrees of freedom in the human motor apparatus, almost every movement can be carried out in an infinite number of ways, with an infinite number of combinations of muscle actions. Thus, while there is great flexibility available to the CNS, the decision as to which joints to move and how much movement is required at each joint presents the CNS with a computational problem of great complexity. "The coordination of a movement," Bernstein wrote, "is the process of mastering redundant degrees of freedom of the moving organ, that is, its conversion to a controllable system."[25] Bernstein asserted that the CNS masters redundant degrees of freedom by linking together two or more degrees of freedom (i.e., muscles or joints) so that they act together as a single unit. In early stages of skill acquisition, this may simply be accomplished by fixation of redundant joints. Later, as greater skill in the motor task is achieved, a more flexible coordination of joint movements is organized, which can be activated and controlled as a whole, rather than as individual commands to the muscles. These coordinations of different joint movements have been referred to as synergies.[26] Note that the idea of movement patterns so prominent in the facilitation model takes on a slightly different meaning here. Movement patterns do not simply result from "wired-in" circuits at low levels of the neural hierarchy. Rather, they result from the problems faced by the CNS as it interacts with the musculoskeletal apparatus and the environment. They are organized by the CNS to solve the redundant degrees of freedom problem. Some patterns may indeed be largely wired in; there is good evidence for this in the case of the locomotor synergy[27] as well as some postural synergies.[28] However, many movement synergies must be learned by building on or combining preexisting synergies.[26]

Another example of the influence of the task-oriented approach has been the application of systems approaches to studying the function of the motor system. Instead of simply studying stimulus–response patterns in motor reflexes, the CNS is viewed as being made up of an enormous number of interacting *feedback* loops.[20] The results of motor output are continually being used to modify and adapt neural output to specific goals and environmental conditions. In addition, the CNS can act in a *feedforward* mode, using sensory input or stored motor programs to guide motor output without the need for feedback.[29]

Finally, the CNS also monitors its own functioning, using internal feedback, so that it can correct errors even before they are expressed in the movement.[30,31]

These ideas present us with a new and rich set of possibilities for exploring the operation of the CNS in controlling movement. The task-oriented or functional perspective that has developed in the neurosciences is exemplified by the shift from an emphasis on motor neurophysiology to an emphasis on the integrated study of motor control. Contemporary motor control research involves a multidisciplinary effort, including neurophysiology, anatomy, muscle physiology, biomechanics, and behavioral sciences. Thus, we are interested in understanding not so much how the CNS produces movement but rather how it achieves *control* of the motor apparatus (i.e., the musculoskeletal system). To achieve this understanding, we must integrate study of the CNS with an understanding of the muscles and joints and the biomechanical principles governing their operation. Furthermore, because movements are organized to solve specific problems or fulfill needs, we must take into account the behavioral level of analysis—how people learn to solve problems in meaningful environments.

The developing motor control perspective that I have been describing has much to offer us in rehabilitation. After all, we must frequently ask ourselves, "What problems does this patient face in controlling his movements to carry out certain tasks?" We are often less interested in the specific site of the neural lesion than in the nature of the control deficit that results. Therefore, the emerging discipline of motor control provides an important source of new assumptions and models on which to begin the process of developing new ways of treating the motor-disabled person.

CRITICAL EVALUATION OF THEORETICAL MODELS UNDERLYING THERAPEUTIC APPROACHES

How do we critically evaluate a therapeutic approach? First, a critique need not necessarily be a destructive process. It is best seen as an analysis, or pulling apart, of the treatment philosophy. This is sometimes painful; it is never easy to examine critically what we do. Nevertheless, it is a necessary first step for improving our methods of treatment. Second, a useful critique of a therapeutic approach should attempt to get to the heart of the assumptions guiding the approach. In other words, we want to understand the deep structure of the approach rather than simply the superficial aspects of the specific techniques. We cannot expect this deep structure to be self-evident. Most of the assumptions underlying a treatment philosophy are implicit and not written down in any organized way. Therefore, in asking what assumptions a particular approach makes, we cannot depend on what is stated but rather we must examine what is actually done. We infer the assumptions based on the way treatments are organized, carried out, and progressed.

To assist in the process of identifying and categorizing the assumptions in particular therapeutic approaches, I will propose a specific framework. This framework allows us to divide the set of assumptions making up a particular therapeutic model into four general categories:

1. *Normal motor control.* What are the assumptions a particular approach makes concerning normal movement control? What neural mechanisms does the approach focus on? The Bobaths, for example, emphasize the importance of the "postural reflex mechanism," including brain stem righting and equilibrium reactions.[32] Proprioceptive neuromuscular facilitation (PNF) places heavy emphasis on the role of the muscle spindle and spinal reflex mechanisms. Both emphasize the importance of proprioceptive and tactile sensation as a prerequisite for movement.
2. *Skill acquisition.* How does a particular approach view development of motor skill both in infancy and in later life? What are the mechanisms by which a person learns movement patterns? What is the proper role of therapist as a teacher of movement? Many of the neurotherapies emphasize the importance of providing the sensation of normal movement. There is also an implicit assumption in many of the therapeutic approaches that mere repetition of a correct movement will lead to increased skill.
3. *Dyscontrol.* How does the treatment approach account for what goes wrong after a brain lesion? How does brain injury interfere with normal movement? Most of the neurotherapies lean heavily on the "release" concept outlined earlier. Most assume that the movement patterns seen after brain injury are "pathological," directly resulting from the lesion.
4. *Recovery of function.* What assumptions does the approach make about neural plasticity and recovery of function after brain injury? What are the neural mechanisms of recovery? Does true restitution occur, or is the observed recovery simply substitution of new behaviors for those that were lost? The principal assumption of the neurotherapies is that recovery of function mimics normal motor development, and an oversimplified developmental sequence is often seen as a model of how recovery should occur.

COMPARISON OF THE FACILITATION MODEL AND THE MOTOR CONTROL MODEL

In this section I will evaluate some of the general assumptions that guide neurophysiological approaches to treatment (the facilitation model) in light of current ideas in the study of motor control (the control model). My purpose here is not to be comprehensive; rather I hope to offer some examples of how this type of analysis can be a useful first step in rethinking our therapeutic strategies.

Normal Movement Control

Assumptions that therapeutic approaches make about the normal mechanisms of movement control are important because they determine the types of movement and the behavioral tasks that are emphasized in rehabilitation. For example, one of the major advances of the neurotherapies was to emphasize the importance of normal postural control and proximal stability for the achievement of distal movement control. This implied that treatment of movement disorders should take into account the component processes, such as posture, that are prerequisites of voluntary movement. The problem, however, with this assumption is that it has been founded on the conception that posture and movement are dependent on reflexes.

Traditional neurophysiology emphasized the experimental study of reflexes as a way of understanding the function of specific neural pathways. Although useful, this approach had the unfortunate effect of promoting a view of voluntary movement as based on combinations of reflexes, some elicited by peripheral stimuli and others elicited by stimuli resulting from the movement itself. This view of movement had a strong influence on the development of the neurotherapies. Many treatment strategies attempt to elicit "normal" reflexes while suppressing or inhibiting abnormal or "pathological" reflexes. Others view all reflexes, normal or not, as a useful way to activate the damaged CNS.

In recent years the importance of reflexes in normal, ongoing movement control has been called into question. Several investigators have now shown that relatively normal patterns of movement are possible even when all sensation from the moving limbs has been eliminated.[33,34] The current view is that most skilled movements are instead dependent on preplanned patterns of neural output to the muscles, referred to as "motor programs."[35] This view is consistent with analysis of the demands placed on humans in most motor tasks. Many skilled movements must take into account changes in the environment that are occurring very quickly. The temporal delays that would be associated with reflexes would make it impossible for a person to accomplish tasks as simple as catching a ball or walking on uneven ground. Most simple movements occur too fast for reflexes to have the time to influence them. Thus, performance of motor tasks in complex environments must be governed by a predictive mode of control rather than a merely reactive one.[36] This implies that a treatment strategy that emphasizes reflexes is not preparing patients to function in the real world.

Even as basic a function as postural control does not and cannot rely on reflexes. For example, when we reach forward to grasp an object, the body's center of gravity also shifts forward. As we reach, we must contract the posterior muscles of the legs and trunk to prevent falling forward. It has now been shown that such postural contractions actually occur before forward sway begins.[37,38] Therefore, these contractions cannot be based on reflexes but rather are preplanned in conjunction with the planning of contractions governing movement of the arm. Effective maintenance of body equilibrium is achieved by predicting the consequences of our movements and precompensating for them, rather than by awaiting the consequences and reacting to them.

Studies of postural responses in humans have shown that even when disturbances of balance occur because of external unexpected stimuli (such as a push or movement of the support surface), the resulting pattern of muscle contractions is not simply a combination of simple reflexes. Instead, a complex synergic pattern of contractions is elicited, the structure of which is critically dependent on the type of disturbance and the behavioral task being performed.[39] The temporal ordering of muscle contractions also differs according to whether the disturbance is imposed from outside (e.g., movement of the floor) or occurs because of voluntary movement (e.g., reaching). Thus, we cannot assume that training a person to react to unexpected disturbances of balance will help that person to learn to precompensate for body sway that results from movements of the limbs.

These ideas also force reexamination of the role of sensation in movement. In general, sensation can function in two different ways: regulatory and adaptive. The regulatory role of sensation is the direct influence it has in shaping an ongoing movement; the adaptive role is the influence it has on succeeding movements. In other words, we can use sensation to guide movements during their execution or we can use it to correct them afterward, so that the next attempt will be better. Ongoing movement must clearly be regulated by exteroceptive sensation, particularly vision, since it must conform to the particular environmental situation. Proprioceptive sensation, on the other hand, probably plays a more adaptive role, allowing us to update our motor programs based on information about how our limbs and muscles are performing as well as where they are at the beginning of the movement.[40]

Most of the neurotherapies have emphasized the particular importance of proprioceptive input in the ongoing control of movement. This is inherent in the use of manual techniques to stimulate certain movements. Thus, patients are learning to react to proprioceptive input rather than learning to use proprioceptive input to improve their movement patterns. In parallel with the overemphasis on the regulatory role of proprioceptive input, there has been insufficient emphasis in the neurotherapies on the importance of patients learning to process visual input to guide their movements.

In summary, then, the neurotherapies assume that proprioceptive reflexes play an important role in normal ongoing movement, especially in the control of posture. This leads to an emphasis in treatment on eliciting reactions rather than on helping the patient to learn to function in a predictive, or feed-forward, mode.

Skill Acquisition

Rehabilitation, for patients, is fundamentally a process of relearning how to move to carry out their needs successfully. The rehabilitation therapist, therefore, must be above all an effective teacher of movement. This implies that we should study and make use of research into the ways people normally learn skilled movement, a field of study referred to as motor learning. The proponents of neurophysiological approaches, by and large, have given this subject little explicit attention in their writings. Thus, we must analyze their assumptions concerning skill acquisition by examining the actual treatment approaches and inferring their assumptions about how skilled movement is developed.

The original aim of the neurotherapies was to develop techniques for inhibiting unwanted patterns of movement and for facilitating normal movements. The idea was that if the patient experienced normal movements, and if these were repeated often enough, the patient would in this way learn to move effectively. This facilitation approach has two problems that I will discuss: (1) it is based on a passive rather than active view of motor learning and (2) it has led to emphasis on learning movement patterns rather than learning to solve motor problems.

The idea that patients can learn movement patterns by experiencing patterns of stimulation that elicit normal movements has its roots in traditional behaviorist psychology, which views learning as establishing associations between particular stimuli and responses. This is essentially a passive view of learning, with the implication that therapists can shape a patient's movement patterns simply by providing appropriate sensory stimulation and reinforcing correct responses. In fact, it is sometimes asserted that conscious effort or participation of the patient is counterproductive in learning, since what is desired is automatic rather than conscious control. As I have already discussed, this approach has simply not been as effective as originally hoped. Although it is sometimes possible to elicit "correct" movements, therapists often find that patients revert to abnormal patterns when the eliciting and reinforcing stimuli are withdrawn.

Modern cognitive psychology takes a more active view of learning. The idea is that patients learn by actively attempting to solve problems. What is learned is not specific solutions but rather general strategies for solving problems. The importance of this approach is evident when we consider the requirements of normal motor learning. It is simply not adequate to learn specific movement patterns because each particular environmental and task situation demands a unique solution (i.e., movement pattern). For example, the act of picking up a cup to drink necessitates very different movement patterns depending on the shape of the cup, how full it is, where it is with reference to the person attempting to pick it up, and what position the person is in (the postural requirements). Thus, it will not help a patient to learn one or a few ways of picking up a cup because these movements will not be of any use in a large number of situations. What must be learned is how to solve the problem of picking up the cup, whatever the specific conditions.

We all recognize that one of the most important prerequisites for motor skill learning is *practice,* but traditionally practice has been viewed simply as repetition of a particular movement. From the cognitive point of view, practice should involve a testing of various strategies in such a way that the person gradually selects and then optimizes the proper control strategy. This might but need not necessarily involve conscious attempts to evaluate performance and alternative strategies. As Bernstein comments,

> The process of practice towards the achievement of new motor habits essentially consists in the gradual success of a search for optimal motor solutions to the appropriate problems. Because of this, practice, when properly undertaken, does not consist in repeating the *means of solution* of a motor problem time after time, but in the *process of solving* this problem again and again by techniques we have changed and perfected from repetition to repetition. It is already apparent here, that, in many cases, "practice is a particular type of repetition without repetition" and that motor training, if this position is ignored, is merely mechanical repetition by rote, a method which has been discredited in pedagogy for some time.[25]

The other problem with the passive view of learning embodied in the neurotherapies is the emphasis on learning movement patterns. This risks, as Gentile has suggested, a confusion as to the goal of movement.[41] If we attempt to teach patients specific movement patterns, we are implicitly teaching them that the goal of movement is to move in a certain way. Although this may be valid in a limited set of situations, such as gymnastics and dance, normally it confuses the means with the end. Bernstein proposed that all purposeful movement is organized to solve specific *motor problems* that arise from the interactions of our needs and desires with the environment. Thus, the specific form of a movement pattern is a means to an end: reaching the goal or solving the motor problem. The correctness of a movement pattern is not determined by whether it corresponds to some ideal form but by whether it is well adapted to the solution of a particular motor problem. The role of the therapist as teacher, therefore, should not simply be to stimulate or facilitate specific movement patterns but rather to select the tasks that are appropriate for the patient to begin to attempt to solve and to structure the relevant conditions of the environment so that patients learn to solve these problems in a variety of contexts.

To summarize, the facilitation model views the patient as a passive recipient of movement patterns; the control model views the patient as an active solver of problems. The role of the therapist, according to the control model, should therefore be to progress the patient through a series of appropriate task situations according to an evaluation of the patient's specific functional problems and his or her progress in overcoming them.

Dyscontrol

How we define the nature of a movement disorder is an important factor determining the treatment approach. The neurotherapies developed on the assumption that in neurological disorders the CNS itself must be the focus of treatment. This assumption was a reaction to the molecular approaches (e.g., muscle reeducation, bracing) that preceded it. Its positive effect was to promote a more aggressive treatment approach: rather than merely trying to compensate for visible deficits, such as paralysis of individual muscles or joint contractures, emphasis was placed on trying to correct the underlying causes of these deficits. With the passage of time, however, we can identify two problems with this approach: (1) overemphasis on positive symptoms and too little attention to the negative symptom and (2) too much reliance on neurophysiological explanations of movement disorders.

The concept of dyscontrol embodied in the neurophysiological approaches to treatment is that the deficit observed after a lesion in the CNS is the result of dissolution. The normal hierarchical "top down" control has been interrupted, and the CNS reverts to a more primitive and automatic level of function. This concept, originally articulated by Jackson, does an excellent job in explaining the *positive* symptoms exhibited after brain damage, such as abnormal movement patterns and spasticity, but it has more difficulty in explaining the *negative* symptoms, such as the specific losses of function that occur. (It also leads to difficulties in explaining how recovery occurs, as will be discussed in the next section.) For example, how do we explain the specific impairments of arm and hand coordination or of gait that occur after stroke? In traditional neurotherapies, the positive symptoms carry most of the explanatory weight. In other words, poor coordination of the arm and hand directly results from spasticity and pathological movement patterns "released" by the CNS lesion; gait disorders are similarly explained. Thus, the dissolution or release assumption logically implies that treatment should be aimed at suppressing the positive symptoms since these are actually causing the observed movement disorders.

The assumption that spasticity is a direct cause of disordered movement has been challenged by a number of researchers in recent years. Sahrmann and Norton showed that spasticity of an antagonist muscle could not explain slowness of movement as is usually assumed, since electromyographic (EMG) recordings of antagonist muscles show that they do not contract with the appropriate timing and magnitude to account for slowness.[42] They proposed that slowness of movement in these cases results directly from insufficient activation of the agonist muscle. Supporting this hypothesis, Tang and Rymer have since demonstrated abnormalities in the EMG output of the weak muscles of hemiplegic patients, indicating that the motor units of the paretic muscles are not activated normally.[43] Perhaps the most telling study was carried out by Nielson and McCaughey.[44] These researchers used a sophisticated biofeedback procedure to train four patients with cerebral palsy to reduce their spasticity. The training process took 18 months, and by the end of this period all four patients were able to practically eliminate spasticity both in the relaxed state and during active contractions. They also used a tracking task to monitor the patients' functional improvement during this time, and, surprisingly, only one of the four patients showed significant improvement in this task, even as they improved their ability to regulate their muscle tone. The patient who did improve in the tracking task showed significant athetosis, and the authors believed that the improvement in function resulted from an improved ability to suppress involuntary movements. Thus, not only are positive symptoms such as spasticity not sufficient as an explanation for observed movement disorders, but we also cannot assume that we will improve function by reducing or eliminating positive symptoms.

The neurotherapies have also placed too much weight on purely neurophysiological explanations for movement disorders. This is not in itself bad, for surely there are neurophysiological explanations. The problem is that it leads us to neglect alternative ways of explaining the deficit that might be more useful for developing treatment strategies. I will discuss two examples.

First, it is usually assumed that the gait disorders associated with spasticity of the lower limb occur because of hyperactive stretch reflexes, especially in the posterior calf muscles. However, Dietz and co-workers, in a series of studies,[45,46] have shown that in stroke patients the EMG of the calf muscles of the affected leg during gait is relatively normal and cannot account for the observed stiffness in the ankle joint. Instead, the increased tone in the ankle joint appears to result from abnormal stiffness of the passive tissues in the ankle joint, that is, contracture. Thus, the observed deficit requires explanation at a biomechanical level, as well as at a neurophysiological level.

Second, the deficit in upper extremity coordination observed in stroke patients is usually attributed to a combination of spasticity and the release of abnormal "synergies," which prevent normal movement patterns. Correspondingly, treatment is aimed at suppressing the unwanted patterns and facilitating individual movements "out of synergy." An alternative way of looking at the deficit, however, is to view the abnormal synergies as learned patterns of movement. Typically, patients with a recent stroke are asked to demonstrate their functional progress by lifting their arm as high as they can. When patients with hemiparesis are given the task of lifting their arm, they do so by using patterns of muscle contraction that are biomechanically advantageous given the current state of their CNS. Gradually, the patients learn these synergies because they help them to accomplish the task (i.e., lifting the arm).

This way of looking at the deficit, as a compensatory strategy, is similar to Taub's concept of "learned nonuse."[47] Taub studied upper extremity function in monkeys subjected to total deafferentation of the limb, and he showed that nonuse of the limb did not occur because of an absolute neural deficit but rather because the monkeys learned to use the intact limb to compensate for the deficits in the deafferented limb. If the intact limb was restrained during the recovery period, or if a bilateral deafferentation had been performed, the deafferented limb was used during functional activities. Taub suggests that learned nonuse may be one of the mechanisms that limit recovery in stroke patients and that preventing such patients from learning to use their unaffected limbs during the initial stages of recovery should help them to achieve a better functional recovery in the long run.

Therefore, the control model views the movement disorder observed after a neural lesion as resulting from compensatory strategies as well as the actual loss of neural tissue. As Grimm and Nashner have commented, "the dimensions of the deficit represent the best mix of systems remaining to participate."[20] Most complex movement skills depend on a number of interacting neural systems; after a focal CNS lesion, there is likely to be considerable "redundancy." In other words, what is observed after a lesion is not just the result of the loss of certain neural tissue but also the result of the attempt of the remaining tissue to compensate for the loss of neural tissue.

One implication of these ideas is that early intervention is extremely important. LeVere has argued that the initial compensatory adaptation of the CNS actually interferes with whatever real recovery might occur later.[48] This is because the compensatory strategies might be "good enough" for patients to accomplish their goals. If so, the patients will not be sufficiently motivated to relearn the movement skills using strategies they used before CNS injury. Thus, it would seem important to intervene early, both to prevent the development of compensatory strategies that will be counterproductive in the long run and to guide the course of recovery so that patients achieve the maximum potential recovery.

Recovery of Function

Current research into the mechanisms involved in neural plasticity following brain damage will be considered in Chapter 4, and therefore I will not discuss this issue in detail. Nevertheless, it is important to point out that our assumptions about the way in which recovery occurs influence both the overall goals we set and the ways in which we progress our treatments.

Ironically, the strict hierarchical model implicit in the facilitation model logically leads to a kind of "therapeutic nihilism," a feeling all too common among therapists and other health care professionals that there can be no true restitution of function after brain damage.[49] If higher centers exert a necessary control over the lower centers, and if damaged neurons in the CNS can never regenerate, then it is difficult to see how any real recovery can occur. In this view, what we call recovery is always substitution of preserved functions for those that were lost. This idea leads to a fatalistic belief that the therapist really has little effect on the course of recovery, beyond prevention of contractures and teaching of alternative strategies.

The more flexible, "distributed" hierarchies that are implicit in motor control theory offer a better way of explaining and understanding the process of recovery.[50] Researchers such as Dichgans and co-workers[51] and Nashner[52] have analyzed recovery strategies and proposed that a kind of "adaptive reweighting" of inputs and outputs occurs after damage to one neural system. These and other studies provide us with the beginnings of a model of recovery that is both optimistic and realistic.

One of the principal assumptions of the facilitation model has always been that recovery of function resembles normal development. This assumption represented an important insight on the part of the founders of the neurotherapies. They observed recovery in their patients and recognized that it had a certain order and regularity, that it followed rules that appeared similar to those in normal development during childhood. Unfortunately, this assumption has often been interpreted too literally; it becomes a recipe for the progression of treatments rather than a guiding model. For example, it is usually assumed that because development proceeds in a proximal to distal order, therapy for upper extremity dysfunction after stroke should be progressed in a similar manner. Control of proximal muscles should be achieved before control of hand function is attempted. This ignores the reality that control of proximal muscles is meaningless to the patient if he or she cannot use the hand. It is the hand that guides the arm rather than the other way around. The proximal to distal sequence is in any case grossly oversimplified in normal motor development. Several studies have shown that infants progressively develop proximal and distal control in parallel rather than in sequential order.[53–55]

It is perhaps more useful to view both normal development and recovery after brain damage as following a common set of rules: those that govern acquisition of skill. The important idea that binds both processes is that motor skill develops out of the need to solve specific motor problems in the environment. Of course, in considering the range of possible solutions, we must take into account the state of the neural machinery as a constraining factor. This is true in both infants and brain-damaged adults. The insight that brain-damaged adults have something in common with developing infants can be useful but should not be used as a blueprint for treatment. Indeed, even if the processes were identical, we know too little about the mechanisms of normal motor development to be so presumptuous as to use it as a recipe for treatment progression.

GENERAL CONSIDERATIONS

Neurophysiological analysis is one way of studying, or observing, the control of movement. We use it to understand the *internal* neural processes responsible for movement. One of the major advantages of the motor control model over the facilitation model is that it promotes a multidimensional view of the motor system, rather than relying primarily on the neurophysiological level of observation. In particular, motor control theory includes two levels of observation that are often not given sufficient emphasis in traditional neurophysiological approaches to treatment. These are (1) biomechanical analysis and (2) behavioral analysis.

A central assumption in the motor control model is that the neural structures that act to control movement must be adapted to constraints imposed by the structure of the musculoskeletal system and by the physical laws governing movement. The biomechanical level of observation views movement from the outside; it includes both study of the mechanical behavior of muscles and kinesiological study of movement of the joints. Analysis of the physical and mechanical problems the motor system must solve will help us to understand the functions of the different neural structures that participate in the control of movement. We can further assume that analysis of the "pathomechanics" and "pathokinesiology" of movement disorders will provide useful insights about the problems that a damaged CNS faces in controlling movement. Pathokinesiology is being increasingly recognized as a field of primary importance to physical therapists.[56,57]

It is also crucial in motor control theory to consider movement from the point of view of the behavioral context in which it occurs. The behavioral level of observation also views movement from the outside. Here, however, the focus is on the outcome of the movement. How successful was the movement in solving the motor problem it was organized to solve? Obviously, an answer to this question requires careful analysis of the functional purpose behind a given motor act, a mode of analysis in which we as therapists need to develop skill. Too often we find ourselves teaching movement skills out of their behavioral context (see Chapter 3). We must also analyze the relevant features of the environment to which the patients must conform their movements. Movements must be shaped to both spatial and temporal features of the environment, such as the shape of a cup we are trying to pick up or the speed of a ball we are trying to catch. As therapists, we exert the most influence on motor learning by our choice of tasks for patients to perform and our structuring of the environments in which they must carry them out.

The important point to be made here is that no one of these levels of observation is enough. If we ignore a particular level, we risk missing useful information about why a patient is having difficulty or about why a treatment is or is not working. For example, manual stabilization of proximal joints may indeed help patients to gain better control of distal movements. The most useful explanation, however, is not necessarily that we have "inhibited" abnormal tone or that we have "facilitated" normal movement patterns. This technique also provides biomechanical stabilization of excess degrees of freedom, which may in turn enable the patient to learn strategies for controlling distal movements that were impossible without external stabilization.

We need to teach ourselves the analytical skills necessary for making a sophisticated biomechanical and functional evaluation of our patients' movement disorders. Furthermore, we should evaluate the efficacy of our treatments not only in terms of their effects on the CNS (i.e., facilitation and inhibition) but also in terms of their effects on the disordered kinesiology and on the ability of the patient to achieve functional outcomes.

In this chapter, I have argued that analysis of the theoretical models underlying our treatment approaches will help us to improve current approaches and to develop new and innovative strategies. The first step in such an analysis must be identification of the general assumptions, both explicit and implicit, that determine the specific strategies we use to treat patients. We should then evaluate a model not in terms of its theoretical purity, or "rightness," but in terms of its usefulness, that is, whether the model generates practical ideas for solving our clinical problems. A historical perspective is certainly useful; it helps us to see that clinical approaches develop out of the practical needs and theoretical biases of the practitioners of the time. Probably the most crucial requirement for establishing an atmosphere conducive to innovation is a willingness to listen to and try new ideas that do not necessarily fit the old models.

I have also offered a framework for analyzing theoretical models, and I have used this framework to compare two general models: the facilitation model and the motor control model. These models, as I have presented them, are certainly oversimplified and by no means comprehensive. My major purpose has been to provide an example of how analysis of theoretical models can be useful.

Finally, it seems reasonable to ask why we even need to evaluate the theoretical assumptions underlying our treatment approaches. If it works, why not just use it? The answer to this question is that without such an analysis, it is difficult to improve what we do. Without an understanding of why our treatments work, we proceed on the basis of a kind of superstition, or pseudoscience. In other words, we are unable to separate the essential aspects of what we did from those aspects that did not help but were originally part of the procedure by chance. Thus, we must continually repeat exactly what we have done without knowing precisely what it was that achieved the beneficial effect. Treatment then becomes ritual. By identifying the theoretical basis of a therapeutic procedure, we can begin to analyze and understand why it works. Only in this way can we improve our procedures for helping patients.

CHAPTER UPDATE (SECOND EDITION)

The preceding portion of this chapter presents a historical perspective on neurological rehabilitation that is, I believe, still valid. Certainly, the neurotherapeutic approaches that were the subject of the analysis are still very much alive, although they are weakened in their influence among clinicians and probably taught more as historical background than as accepted practice in many schools of physical therapy.[58] Among many developments in the 12 years since this chapter was first published, models of *disablement* have come into considerable prominence in the physical therapy literature.[59,60] Disablement models can be useful for both explaining and predicting disability, as well as for analyzing the role of various therapeutic approaches. Further, these models share many assumptions with the ideas expressed in this and other chapters in this book. Therefore, I thought it would be useful to add some discussion of the disablement concept to this chapter.

THE NAGI MODEL OF DISABLEMENT

The concept of disablement, especially as put forward by Nagi[61] and Jette,[62] is a way of explaining or understanding the outcomes of disease processes. In this view, the consequences of pathology or disease can be classified into four different levels, producing a casual chain from pathology to disability (see Figure 1–1A). Other classification systems, such as those put forward by the World Health Organization, are based on the identical concept of disablement, but define the levels in slightly different ways.[63] For simplicity, this discussion will consider the model put forward by Nagi, but the conclusions are also relevant to the other models.

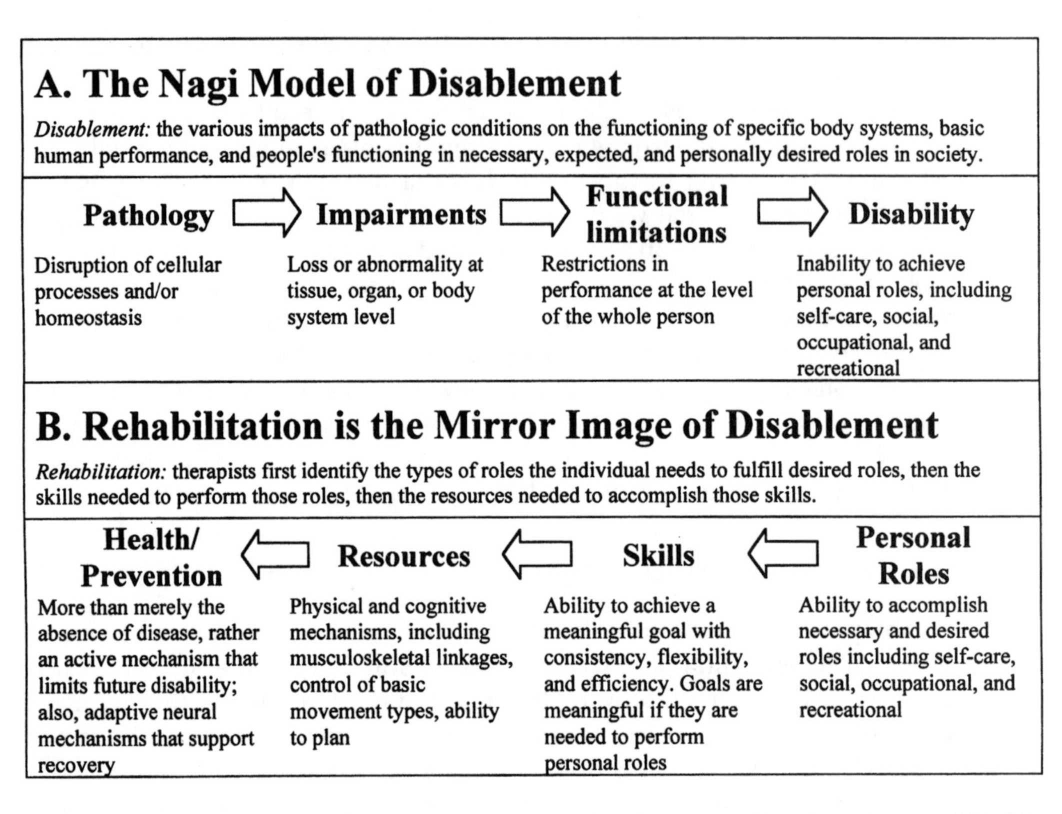

Figure 1–1 A. The Nagi model of disablement. B. Rehabilitation is the mirror image of disablement.

In the Nagi model, a blood clot that causes infarction of neurons in a specific part of the cerebral cortex, a process referred to as stroke, is an example of a *pathology*. The pathology may ultimately lead to *disability*, defined as the inability of the individual to accomplish his or her personal roles, including self-care, social, occupational, and recreational roles.

Prevention and rehabilitation of disability are what physical therapists do. But, with few exceptions, physical therapists do not treat the patient's pathology or disability directly. Instead, they typically intervene either by remediation of basic motor control mechanisms or by the training of the patient in functional skills. That is, physical therapy takes place—for the most part—at the interface between impairments and functional limitations, where *impairment* refers to lost or abnormal functioning of a tissue, organ, or body system and *functional limitations* are losses or abnormalities at the level of the whole person. For example, the death of neurons associated with stroke may lead to impairments such as weakness, spasticity, and incoordination, and these impairments may in turn lead to certain functional limitations, such as inability to get dressed, inability to drive, and inability to walk from the bedroom to the bathroom. The functional limitations may then lead to specific disabilities, such as inability to take care of oneself independently, to work at one's regular job, and to accomplish one's regular family and social roles.

Although the arrows between levels in Figure 1–1 appear to imply direct causal relationships, no obligatory associations exist between levels. Weakness of the quadriceps muscle does not necessarily impair the ability to walk up and down stairs. An inability to walk on stairs does not necessarily prevent an individual from carrying out occupational activities. The disablement framework strongly implies that a specific impairment, such as weakness, should be a focus of intervention *if and only if* that impairment can be demonstrated to be a causal factor in the inability to carry out a functional activity that is necessary for the individual in his or her life activities. This emphasis on *functional validity* as the measure of effectiveness of a therapeutic approach is perhaps the most important contribution of the disablement concept as a prescriptive model.

The next logical step, then, would be to measure our success in therapeutic practice in functional dimensions. Here, the disablement model can serve not just as a prescriptive model, but also as a tool. The various disablement models provide classification systems that should—in theory—facilitate the investigation of the causal links between specific impairments and functional limitations. Unfortunately, the classification systems and the definitions upon which they are based remain somewhat vague. In particular, there is often confusion in the way that functional activities are defined and classified.

The defining characteristic of a functional activity is that it has a goal and context that are meaningful to the individual. Achieving a certain magnitude of knee or ankle flexion during the swing phase of gait is not a meaningful goal to the patient. Getting from bed to bathroom in the patient's home is a meaningful goal, and it can, in fact, be accomplished using many different gait patterns, some more effective and economical than others. The relationship of impairments to functional limitations is similar conceptually to the means-end relationship between movements and actions, as described by Gentile in Chapter 3. Thus, an abnormal or dysfunctional gait pattern is an impairment; an inability to get from the living room to the kitchen is a functional limitation. Balance is another motor control mechanism that is sometimes classified as a functional activity. Although adequate balance is critical in many functional activities (eg, dressing, eating), balance deficits should be classified as impairments, not functional limitations. Similarly, reaching, grasping, manipulation, and many other types of movement patterns are elements of functional activities; they are not functional activities *per se*.

One useful outcome of attempting to analyze current physical therapy practice through the lens of a disablement model is that we can conclude that our nomenclatures for describing and classifying impairments and functional limitations are in need of a major overhaul. In particular, the descriptive classification of impairments currently in use for neurological patients was designed by neurologists primarily for diagnosing the nature and location of the pathology. Thus, changes in reflexes or tone, upper and lower motor neuron patterns of weaknesses, and similar impairments have considerable value for helping the physician to make a medical diagnosis, and they may have some prognostic value. However, physical therapists need a system of describing impairments that will help to clarify the causes of functional limitations and how to remediate them.[20]

A more functionally useful classification of impairments might be adapted from research in motor control, as a "guiding set of assumptions." In particular, it is increasingly evident that basic movement patterns such as walking,[64,65] balance,[66] reaching,[67,68] and precision grip[67,69–71] are controlled by defined neural systems and can be selectively impaired. Knowledge of the organization of these movements and the rules by which they are controlled can presumably help one to understand better how to remediate the functional problems that occur when these movement types are impaired. At least two recent textbooks addressed to rehabilitation therapists use such a motor control framework to analyze the relationships between impairments and functional limitations and to suggest possible interventions.[72,73]

Although motor control research is moving in the direction of developing more functionally useful categories of movement control, considerable work needs to be done, especially with individuals of different ages and with different pathologies. An exemplary approach can be seen in the work of Patla, who is systematically examining the cognitive and motor mechanisms used by patients to navigate and avoid obstacles in complex environments.[74,75] We may think of this as a "top-down approach," in which the investigator first defines a functional activity and then analyzes the mechanisms used by the individual to accomplish the goal.

Rehabilitation is the Mirror Image of Disablement

The notion of a top-down approach may also be useful in applying disablement frameworks, such as the Nagi model, to clinical practice. Indeed, rehabilitation may be thought of as the mirror image of the disablement process (see Figure 1–1B). In this view, the causal arrows are in the opposite direction, and the levels are expressed in positive rather than negative terms. Physical therapy intervention begins with a specification of the desired end result, in terms of the personal and social roles the patient is attempting to achieve or resume or keep from losing. The therapist then determines which functional skills are needed by the individual to accomplish these specific roles. If certain of these functional skills need improvement, the therapist may choose to enhance, by exercise or by practice, certain motor control mechanisms, such as strength of a muscle group, control of movement velocity, or ability to link grip aperture to object size in reaching tasks. These mechanisms can be thought of as resources that are used, often flexibly, in the performance of the functional skills. Thus, therapeutic exercise is justified when it is designed to enhance resources that are needed for the performance of critical functional skills.

In this rehabilitation model, the opposite of pathology is health. Health is not simply the absence of disease, but an active process of healing.[76] In many instances, especially in the acute stages of a pathological process, the primary goal of therapy is to create an optimal environment for tissue healing and system reorganization. Understanding the different types of recovery processes and the factors that influence them, as discussed by Held in Chapter 4, is essential to maximizing health as an outcome in the rehabilitation process.

Measuring Functional Outcomes—Beyond Independence

If the outcome of physical therapy for neurological patients is the learning or relearning of critical skills, then we need tools to measure the degree to which patients are successful in acquiring skill. Unfortunately, most of our tools for measuring functional performance focus heavily, if not exclusively, on the amount of assistance required. Independence in a functional task means simply that the performer can accomplish the task without assistance. It does not ensure that the task is performed skillfully, nor does it ensure that the task is performed well enough so that it will be useful to the patient in the context in which it will be carried out. In fact, specification of the amount of assistance needed in functional tasks is more a measure of disability (that is, how well the individual accomplishes their personal roles) than functional outcome.

A more useful approach is to measure performance in functional activities in terms of the skill with which they are performed. *Skill is the ability to achieve a goal consistently, flexibly, and efficiently.* This simple definition provides a set of dimensions within which functional outcomes can be measured: degree of goal attainment, consistency of goal attainment, environments and contexts within which the goal can be attained, physical efficiency (endurance), and cognitive efficiency (attention required).

CONCLUSION

What the physical therapist does in neurological rehabilitation is to manipulate the tasks that a patient will practice. In other words, the task, or what Bernstein referred to as the motor problem,[25] is the fundamental unit of therapy. The therapist selects the tasks, structures the environment and context for the task, varies certain task parameters, progresses the subject to more challenging tasks, and switches to new tasks when the patient is ready. This involves, as described by Gentile in Chapter 3, a complex process of task analysis that every therapist, as a teacher of movement skills, must engage in. This is the real clinical skill of the therapist, more so than the manual skills that are typically emphasized, especially by younger therapists. In Chapter 2, Carr and Shepherd describe with specific examples the application of such a task-analytic approach to the rehabilitation of neurological patients.

Disablement and rehabilitation models are the *yin* and *yang* of physical therapy intervention. They reflect complementary ways of understanding what physical therapists do. Disablement reflects a medical model, in that it provides a way of analyzing what is "wrong" with the patient. Such an approach often leads to a determination of what the therapist should do to treat the patient. Rehabilitation reflects a patient-centered approach in which the focus is on the patient's goals and what he or she needs to do to reach them. In this view the patient plays an active role in the rehabilitation process. Each approach has its advantages and limitations, and the right balance very much depends on the individual situation.

REFERENCES

1. Knott M, Voss DE. *Proprioceptive Neuromuscular Facilitation.* 2nd ed. New York: Harper & Row; 1968.
2. Bobath B. *Adult Hemiplegia: Evaluation and Treatment.* 2nd ed. London: Heinemann; 1978.
3. Rood MS. Neurophysiological mechanisms utilized in the treatment of neuromuscular dysfunction. *Am J Occup Ther.* 1956;10:220–225.
4. Brunnstrom S. *Movement Therapy in Hemiplegia.* New York: Harper & Row; 1970.
5. Fay T. The use of pathological and unlocking reflexes in the rehabilitation of spastics. *Am J Phys Med.* 1954;33:347–352.
6. Bouman HD. Proceedings: An exploratory and analytical survey of therapeutic exercise (Northwestern University Special Therapeutic Exercise Project—NUSTEP, July 25, 1966 to August 19, 1966). *Am J Phys Med.* 1967;46(1).
7. Payton OD, Hirt S, Newton RA, eds. *Scientific Bases for Neurophysiologic Approaches to Therapeutic Exercise: An Anthology.* Philadelphia: FA Davis; 1977.
8. Hirt S. Historical bases for therapeutic exercise. *Am J Phys Med.* 1967;46:32–38.
9. Pohl JF. *The Kenny Concept of Infantile Paralysis and Its Treatment.* Minneapolis-St. Paul: Bruce; 1943:151–152.
10. Beard G. Foundations for growth: a review of the first forty years in terms of education, practice, and research. *Phys Ther Rev.* 1961;41:843–861.
11. Pinkston D, ed. Analysis of traditional regimens of therapeutic exercise. *Am J Phys Med.* 1967;46:713–731.
12. Treanor WJ, Cole OM, Dabato R. Selective reeducation and the use of assistive devices. *Phys Ther Rev.* 1954;34:618–625.
13. Levitt S. Physiotherapy in cerebral palsy today. *Phys Ther Rev.* 1955;35:430–437.

14. Bobath K, Bobath B. The neuro-developmental treatment of cerebral palsy. *Phys Ther.* 1967;47:1039–1041.
15. Walshe FMR. Contributions of John Hughlings Jackson to neurology: a brief introduction to his teachings. *Arch Neurol.* 1961;5:119–131.
16. Evarts EV. Representation of movements and muscles by pyramidal tract neurones of the precentral motor cortex. In: Yahr MD, Purpura DP, eds. *Neurophysiological Basis of Normal and Abnormal Motor Activities.* New York: Raven Press; 1967:215–253.
17. Semans S. The Bobath concept in treatment of neurological disorders. *Am J Phys Med.* 1967;46:732–785.
18. Perry CE. Principles and techniques of the Brunnstrom approach to the treatment of hemiplegia. *Am J Phys Med.* 1967;46:789–812.
19. Beevor CE. The Croonian lectures on muscular movements and their representation in the central nervous system (originally published in British Medical Journal, 1903) Excerpted in: Payton OD, Hirt S, Newton, RA, eds. *Scientific Bases for Neurophysiologic Approaches to Therapeutic Exercise: An Anthology.* Philadelphia: FA Davis, 1977.
20. Grimm, RJ, Nashner LM. Long loop dyscontrol. In: Desmedt JE, ed. *Progress in Clinical Neurophysiology, Vol 4, Cerebral Motor Control in Man: Long Loop Mechanisms.* Basel, Switzerland: Karger; 1978:70–84.
21. Carr JH, Shepherd RB. *A Motor Relearning Programme for Stroke.* London: Heinemann; 1982.
22. Kuhn TS. *The Structure of Scientific Revolutions.* 2nd ed. Chicago: University of Chicago Press; 1970.
23. Greene PH. Problems of organization of motor systems. In: Rosen R, Snell FM, eds. *Progress in Theoretical Biology.* New York: Academic Press; 1972:303–338.
24. Prochazka A, Hulliger M. Muscle afferent function and its significance for motor control mechanisms during voluntary movements in cat, monkey, and man. In: Desmedt JE, ed. *Motor Control Mechanisms in Health and Disease, Vol 39, Advances in Neurology.* New York: Raven Press; 1983:93–132.
25. Bernstein NA. *The Coordination and Regulation of Movements.* New York: Pergamon; 1967:127, 134.
26. Gelfand IM, Gurfinkel VS, Tsetlin ML, Shik ML. Some problems in the analysis of movement. In: Gelfand IM, Gurfinkel VS, Fomin SV, Tsetlin ML, eds. *Models of the Structural-functional Organization of Certain Biological Systems.* Cambridge, MA: M.I.T. Press; 1971:329–345.
27. Forssberg H. Ontogeny of human locomotor control: I. Infant stepping, supported locomotion and transition to independent locomotion. *Exp Brain Res.* 1985;57:480–493.
28. Gurfinkel VS, Kots YM, Paltsev EI, Feldman AG. The compensation of respiratory disturbances of the erect posture of man as an example of interarticular interaction. In: Gelfand IM, Gurfinkel VS, Fomin SV, Tsetlin ML, eds. *Models of the Structural-functional Organization of Certain Biological Systems.* Cambridge, MA: M.I.T. Press; 1971:382–395.
29. Kelso JAS. Concepts and issues in human motor behavior: coming to grips with the jargon. In: Kelso JAS, ed. *Human Motor Behavior: An Introduction.* Hillsdale, NJ: Lawrence Erlbaum Associates; 1982:21–61.
30. Higgins JR, Angel RW. Correction of tracking errors without sensory feedback. *J Exp Psychol.* 1970;84:412–416.
31. Evarts EV. Feedback and corollary discharge: a merging of the concepts. *Neurosci Res Prog Bull.* 1971;9:86–112.
32. Bobath K. The normal postural reflex mechanism and deviation in children with cerebral palsy. *Physiotherapy.* 1971;57:515–525.
33. Polit A, Bizzi E. Characteristics of motor programs underlying arm movements in monkeys. *J Neurophysiol.* 1979;42:183–194.
34. Taub E, Berman AJ. Movement and learning in the absence of sensory feedback. In: Freedman SJ, ed. *The Neuropsychology of Spatially Oriented Behavior.* Homewood, IL: Dorsey; 1968:173–192.
35. Keele SW. Movement control in skilled motor performance. *Psychol Bull.* 1968;70:387–403.
36. Poulton EC. On prediction in skilled movements. *Psychol Bull.* 1957;54:467–478.
37. Belenkii VY, Gurfinkel VS, Paltsev YI. Elements of control of voluntary movements. *Biophysics.* 1967;12:154–161.

38. Cordo PJ, Nashner LM. Properties of postural adjustments associated with rapid arm movements. *J Neurophysiol.* 1982;47:287–302.
39. Nashner LM, McCollum G. The organization of human postural movements: a formal basis and experimental synthesis. *Behav Brain Sci.* 1985;8:135–172.
40. Bizzi E. Central and peripheral mechanisms in motor control. In: Stelmach GE, Requin J, eds. *Tutorials in Motor Behavior.* Amsterdam: North-Holland; 1980:131–143.
41. Gentile AM. A working model of skill acquisition with application to teaching. *Quest.* 1972;44:1–11.
42. Sahrmann SA, Norton BJ. The relationship of voluntary movement to spasticity in the upper motor neuron syndrome. *Ann Neurol.* 1977;2:460–465.
43. Tang A, Rymer W. Abnormal force-EMG relations in paretic limbs of hemiparetic human subjects. *J Neurol Neurosurg Psychiatry.* 1981;44:690–698.
44. Nielson PD, McCaughey J. Self-regulation of spasm and spasticity in cerebral palsy. *J Neurol Neurosurg Psychiatry.* 1982;45:320–330.
45. Dietz V, Quintern J, Berger W. Electrophysiological studies of gait in spasticity and rigidity: evidence that mechanical properties of muscles contribute to hypertonia. *Brain.* 1981;104:431–449.
46. Dietz V, Berger W. Normal and impaired regulation of muscle stiffness in gait: a new hypothesis about muscle hypertonia. *Exp Neurol.* 1983;79:680–687.
47. Taub E. Somatosensory deafferentation research with monkey: implications for rehabilitation medicine. In: Ince LP, ed. *Behavioral Psychology in Rehabilitation Medicine: Clinical Implications.* Baltimore: Williams & Wilkins; 1980:371–401.
48. LeVere TE. Recovery of function after brain damage: a theory of the behavioral deficit. *Physiol Psychol.* 1980;8:297–308.
49. Walsh RN, Cummins RA. Neural responses to therapeutic sensory environments. In: Walsh R, Greenough W, eds. *Environments as Therapy for Brain Dysfunction, Vol 17, Advances in Behavioral Biology.* New York: Plenum Press; 1976:171–200.
50. Arbib MA. Perceptual structures and distributed motor control. In: Brooks VB, ed. *Handbook of Physiology, Sec 1, The Nervous System, Vol 2, Motor Control, Part 2.* Bethesda, MD: American Physiological Society; 1981:1448–1480.
51. Dichgans J, Bizzi E, Morasso P, Tagliasco V. Mechanisms underlying recovery of eye-hand coordination following bilateral labyrinthectomy in monkeys. *Exp Brain Res.* 1973;18:548–562.
52. Nashner LM. Analysis of movement control in man using the movable platform. In: Desmedt JE, ed. *Motor Control Mechanisms in Health and Disease, Vol 39, Advances in Neurology.* New York: Raven Press; 1983:607–619.
53. Halverson HM. An experimental study of prehension in infants by means of systematic cinema records. *Genet Psychol Monogr.* 1931;10:18–40.
54. Bruner JS. Organization of early skilled action. *Child Dev.* 1973;44:1–11.
55. Hofsten C von. Development of visually directed reaching: the approach phase. *J Hum Movement Stud.* 1979;5:160–178.
56. Hislop HJ. Tenth Mary McMillan Lecture: the not-so-impossible dream. *Phys Ther.* 1975;55:1069–1080.
57. Rothstein JM. Pathokinesiology—A name for our times? *Phys Ther.* 1986;66:364–365.
58. Riolo, L, ed. Special issue: education. *Neurology Report.* 1996;20(1):9–63.
59. American Physical Therapy Association. Guide to Physical Therapist Practice. *Phys Ther,* 1997;77:1163–1650.
60. Jette A, ed. Special issue: physical disability. *Phys Ther,* 1994;74:379–503.
61. Nagi S. Some conceptual issues in disability and rehabilitation. In: Sussman M, ed. *Sociology and Rehabilitation.* Washington, DC: American Sociological Association; 1965:100–113.
62. Jette, AM. Physical disablement concepts for physical therapy research and practice. *Phys Ther,* 1994;74:380–386.

63. *International Classification of Impairments, Disabilities, and Handicaps*. Geneva, Switzerland: World Health Organization; 1980.
64. Gordon J. Spinal mechanisms of motor coordination. In: Kandel ER, Schwartz JH, Jessell TM, ed. *Principles of Neural Science*. New York: Elsevier; 1991:581–595.
65. Winter, D. Coordination of motor tasks in human gait. In: SA Wallace, eds., *Perspectives on the Coordination of Movement (Advances in Psychology, Vol. 61)*. New York: Elsevier; 1989:329–363.
66. Ghez C. Posture. In: Kandel ER, Schwartz JH, Jessell TM, ed. *Principles of Neural Science*. New York: Elsevier; 1991:596–607.
67. Jeannerod, M. *The Neural and Behavioral Organization of Goal-Directed Movements*. Oxford: Clarendon Press; 1988.
68. Georgopolous, AP. On reaching. *Ann. Rev. Neurosci.* 1986;9:147–170.
69. Johansson RS, Riso R, Häger C, Bäckström L. Somatosensory control of precision grip during unpredictable pulling loads: I. Changes in load force amplitude. *Exp. Brain Res.* 1992;89:181–191.
70. Johansson RS, Häger C, Riso R. Somatosensory control of precision grip during unpredictable pulling loads: II. Changes in load force rate. *Exp. Brain Res.* 1992;89:192–203.
71. Johansson RS, Häger C, Bäckström L. Somatosensory control of precision grip during unpredictable pulling loads: III. Impairments during digital anesthesia. *Exp. Brain Res.* 1992;89:204–213.
72. Shumway-Cook A, Woollacott MH. *Motor Control: Theory and Practical Applications*. Baltimore, MD: Williams & Wilkins; 1995.
73. Carr J, Shepherd R. *Neurological Rehabilitation: Optimizing Motor Performance*. Oxford, England: Butterworth-Heinemann; 1998.
74. Patla AE, ed. *Adaptability of Human Gait: Implications for the Control of Locomotion*. Amsterdam: North-Holland; 1991.
75. Patla AE. Age-related changes in visually guided locomotion over different terrains: major issues. In: Stelmach G, Homberg V, eds. *Sensorimotor Impairment in the Elderly*. Dordrecht, The Netherlands: Kluwer; 1993: 231–252.
76. Rimmer JH. Health promotion for people with disabilities: the emerging paradigm shift from disability prevention to prevention of secondary conditions. *Phys Ther.* 1999;79:495–502.

A Motor Learning Model for Rehabilitation

Janet Carr and Roberta Shepherd

INTRODUCTION

The individual with movement dysfunction associated with a brain lesion needs to learn again how to perform the actions of daily life as effectively as possible given the nature of the lesion. The emphasis in movement rehabilitation, we first argued nearly 2 decades ago, should be on training the individual to optimize functional motor performance rather than on giving therapy to bring about a localized physiological change that is assumed to carry over automatically into improved function. Our point was, that in terms of improving motor performance, a learning or relearning model may be a more appropriate approach for rehabilitation than the common (at that time) neurofacilitation model. We pointed out that an emphasis on task-specific exercise and training would avoid the problem of transfer into real life and enable the individual to improve effectiveness of motor performance, and it would minimize the development of the secondary problems related to lesion-induced disuse. The neurofacilitation model is still in common use in the 1990s despite both a lack of evidence of its effectiveness and of a clear and scientifically valid theoretical rationale. It is puzzling that there is a reluctance in physical therapy to embrace new methodologies that are theoretically sound and shown to have a positive effect on function.

We would argue that clinical practice can, increasingly, be based on evidence, ie, on clinical and experimental studies. Increasing evidence indicates the effectiveness of a dynamic, more task-related rehabilitation process with emphasis on the need for the patient to exercise and practice to increase strength, fitness, and skill. Such evidence is provided by different levels of investigation ranging from the randomized controlled trial to the formalized single case study. On the whole, these studies illustrate the process of developing clinically testable hypotheses out of investigations of impairments following brain lesion, the characteristics of effective motor performance, skill acquisition, and the effects of exercise. Relevant studies of stroke patients include:

- Training reaching in sitting[1]
- Training sit-to-stand[2–4]
- Training walking with a treadmill[5,6]
- Effects on gait of lower limb exercise and walking practice[7]
- Training sensory perception[8,9]
- Training upper limb function using constraint of unaffected limb plus forced use of the hemiparetic limb[10,11]
- Training balanced movement in standing with harness support[12]

In addition, it has been found that:

- Intensive upper limb exercise led to improved recovery of arm function[13]
- Reduction of energy expenditure and cardiovascular demands followed treadmill aerobic exercise[14]
- Casting and stretching increased passive ankle dorsiflexion following traumatic brain injury[15]

Evidence also shows that strengthening exercise for the lower limbs increased walking speed and decreased the incidence of falls in elderly people,[16] and increased strength and improved functional performance in children with cerebral palsy.[17,18]

Where it has been generally considered self-evident by clinicians that rehabilitation enables the individual to "make the most of" remaining neural systems, it is becoming clear that what an individual experiences and practices is likely to drive the establishment of new functional neural connections. It is increasingly being understood that central and peripheral neural mechanisms, and the muscular and cardiovascular systems are highly adaptive, responsive to both internal and external processes, and that use and experience, whether rich or impoverished, have anatomical, physiological, and behavioral sequelae. Rehabilitation aims to ensure, therefore, that negative adaptive changes related to disuse, immobility, and a reduced lifestyle (eg, stiff short muscles, secondary muscle weakness, reduced work capacity, poverty of experience with reduced physical, mental, and social interaction) do not take place or are minimized following an acute brain lesion. Rehabilitation is directed at ensuring that rehabilitation is a positive experience for the patient.

This chapter provides a review of a scientific rationale for movement rehabilitation, outlined with reference to the movement dysfunction resulting from acute brain lesion such as stroke. It is in this situation that one sees rather clearly the need to create a motivating and active environment in which the person who is motor-disabled has the opportunity to learn again how to move with maximum efficiency and effectiveness. The chapter includes additional emphasis on training and exercise designed to strengthen muscles, to preserve soft tissue length, to improve motor control, and to build up endurance and physical fitness. We argue for the need to increase the time spent in daily practice and exercise by emphasizing group work, including circuit training around work stations with therapists supervising a group of patients and assisting individuals where necessary. The section on impairments underlying functional disability reflects the considerable body of information now available on the mechanisms of impairments and makes a distinction between primary and secondary mechanisms.

To train people with disabilities to improve performance and regain skill in actions of daily life, the physical therapist needs a description of the biomechanical features critical to effective performance of an action, an understanding of motor control and skill acquisition, an understanding of the underlying processes of impairment, and an understanding of the negative and positive effects on brain organization of the patient's experience. From such knowledge emerge principles of exercise prescription and training, enabling the refinement of old and development of new methodologies to improve muscle strength, motor performance, and the flexibility and adaptability that are necessary for skilled action in the complex environments of daily life.

In the past decade there has been a considerable increase in investigations of the biomechanics of actions performed by able-bodied individuals and other studies describing the major features of disabled performance, and these studies provide critical information for the design of training programs. This information improves our understanding of what constitutes effective (ie, skilled) performance of particular actions and enables us to build models against which we can compare the performance of the person with a disability before and after intervention as a guide to training and exercise.

Description of the action involves more than a simple statement of the end result. For example, a description of reaching for an object while standing includes eye movement, angular (segmental) displacements of upper body and lower limbs as well as of the arm(s), muscle forces and postural adjustments, factors such as preshaping of the hand before grasping the object, and speed changes from the faster reaching phase to the relatively slower terminal phase as the hand approaches the object (Figure 2–1).[19]

An understanding of the critical part played by visual inputs concerning the spatial and textual properties of the object and by tactile inputs concerning the object's characteristics once it is grasped is also necessary. The therapist needs an understanding of the muscle activity involved (prime movers or synergists, concentric or eccentric modes), the relative timing of muscle onsets and segmental rotations, and the relative timing of muscle activity involved in postural adjustments occurring in the lower limbs and trunk before the onset of the reaching movement.

Critical to the understanding of movement is knowledge of the interactions between limbs and segments. It is rare that one part of the body functions

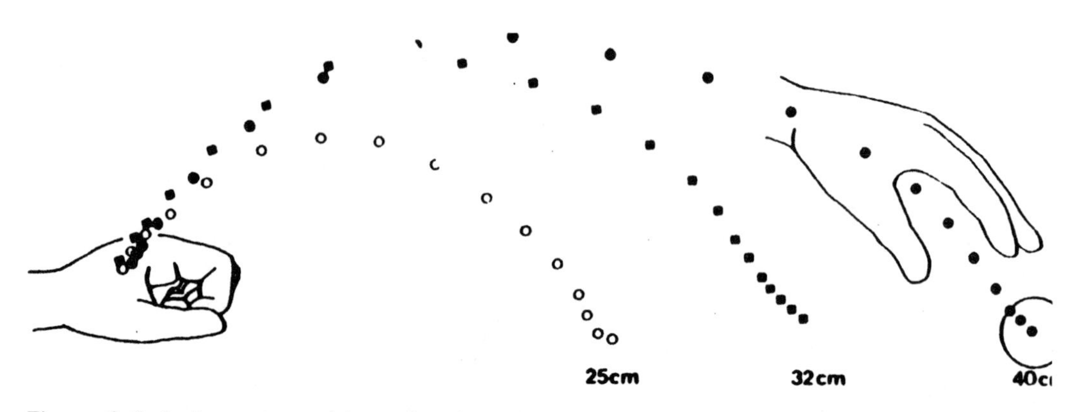

Figure 2–1 A cinematographic study of reaching to grasp an object. The traces illustrate the linear velocity of the hand. The hand slows as it nears the object. Other research shows that aperture formation starts when movement starts, reaches a maximum around the same time as the movement of the hand slows, and decreases until it is the size necessary for grasping. *Source:* M. Jeannerod, The Timing of Natural Prehension Movements, *Journal of Motor Behavior*, Vol. 16, p. 239, 1984. Reprinted with permission of the Helen Dwight Reid Educational Foundation. Published by Heldref Publications, 1319 Eighteenth St., N.W., Washington, D.C. 20036-1802. Copyright © 1984.

in isolation from other parts. The linked-segment dynamics enabling our actions to be effective are complex. Their complexity is such that the regaining of skilled motor performance in individuals with disabilities demands practice of specific actions in their natural contexts. Two recent studies exemplify these complex interactions but many other biomechanical studies demonstrate action-specific dynamics. Diener and colleagues[82] examined subjects rising on tiptoe in standing. Initially, that is, before heel raise, the body mass moves forward with activation of tibialis anterior (TA), quadriceps (QUAD), and biceps femoris (BF). TA action brings body mass forward, and QUAD and BF stabilize the knee because the knee will be affected by the upcoming calf muscle action. These findings are illustrations that timing of segmental movement and muscle activity is crucial to successful performance. In a study of standing in which subjects were asked to shift their weight on to one leg either sideways or forward, Goldie and colleagues' findings illustrated the cooperative interaction between the lower limbs in this simple action and how that cooperation breaks down in the presence of impairments following stroke.[81] Shifting body weight in either plane involves complex dynamics in which a propulsive force is produced by one leg and a braking force by the other, with hip abductor and adductor muscles in particular coupled to produce the action. Dynamic interactions between the extensor muscles of the lower limbs are also critical to preserving the supportive function of the limbs throughout the action. The results of this second study highlight the need in clinical practice to train weight shifting to *both* legs. Both studies illustrate the significance of biomechanical information to the understanding of movement control, the analysis of movement, and the planning of intervention.

The transfer of learning from rehabilitation environment to the world outside is a critical issue if movement rehabilitation is to be functionally effective as well as physiologically and economically efficient. Training involves the practice of those actions in which skilled performance is required, and exercises to improve strength and control should incorporate similar components to those that are critical to performance of these actions. Given the large number of actions to be trained, guidance in organizing a training program comes from research indicating the simplifying strategies used normally by the system to control the many degrees of freedom inherent in linked segment dynamics. From research into biomechanics and motor control mechanisms comes information that enables us to classify into groups those actions that are basically, ie, mechanically, similar. For example, actions involving the lower limbs, most of which take place with one or both feet on the support surface, require extension and flexion of lower limb segments over a fixed foot (feet) (Figure 2–2) with body mass balanced throughout the action, eg, sit-to-stand, stair climbing, stance phase of walking, bending to pick up an object from the floor in standing. Manipulative actions are similarly carried out by using a (relatively) small number of finger groupings[20] (Figure 2–3). From such information, it is possible to plan exercises that are likely to transfer into improved performance of similar actions.

It has been argued[21] that if training to improve effectiveness of an action is the major goal of movement rehabilitation, then the most relevant way of testing outcome is to measure performance of that action to see whether it is more effective and requires less energy. Performance is measured by functional scales, time taken in performing the action, and accuracy in achieving the goal. Biomechanical instrumentation is used to determine kinematic details such as path of body parts (eg, the path of the shoulder when standing up), angular displacement of joints, and kinetic details about muscle forces and ground reaction forces. Energy expenditure is measured by testing of physiological responses such as heart rate and oxygen uptake.

Training to improve skill (any activity that has become better organized and more effective as a result of practice[22]) requires an understanding of the relationship between input from the environment and the control of goal-directed intentional movement as well as an understanding of the active processes involved in detection of and attention to relevant cues from the environment. Cognition needs action and action itself is part of cognition. Physical therapists are sometimes concerned that emphasis on cognitive processes in training does not consider those people whose cognitive function is impaired. Nevertheless, the link between cognition and action cannot be ignored and action is unlikely to improve without training that includes matching intention to action. Individuals with memory deficits, for example, require

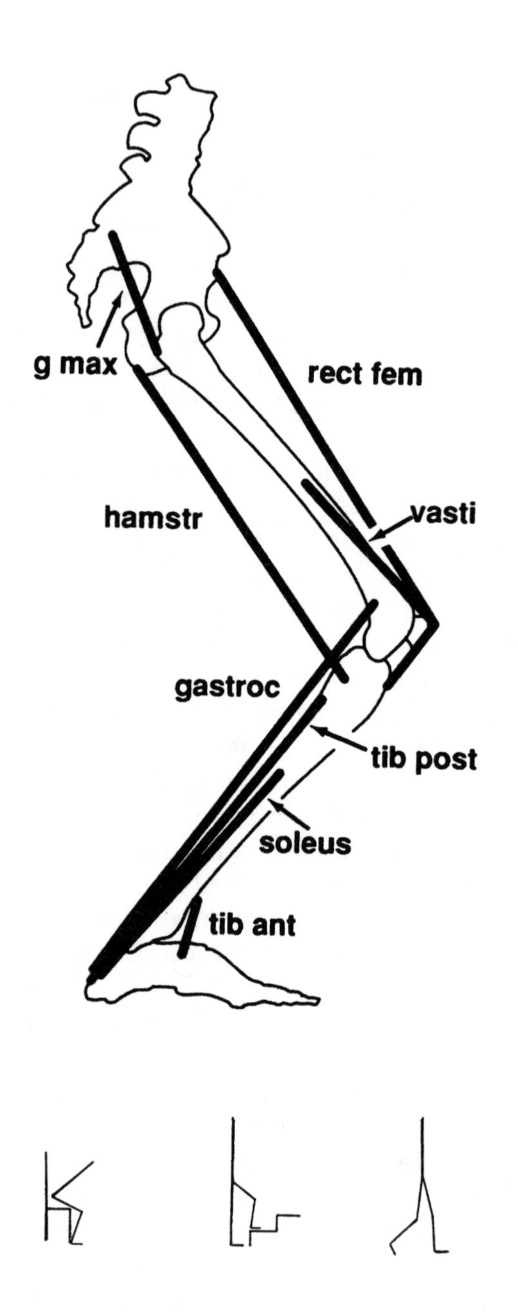

Figure 2–2 Diagram showing lower limb segments and eight muscle groups affecting segmental rotation during flexion and extension of lower limbs over a fixed foot. *Source:* Reprinted from *Journal of Biomechanics*, Vol. 26, A.D. Kuo and F.E. Zajac, A Biomechanical Analysis of Muscle Strength as a Limiting Factor in Standing Posture, p. 139, © 1993, with permission from Elsevier Science.

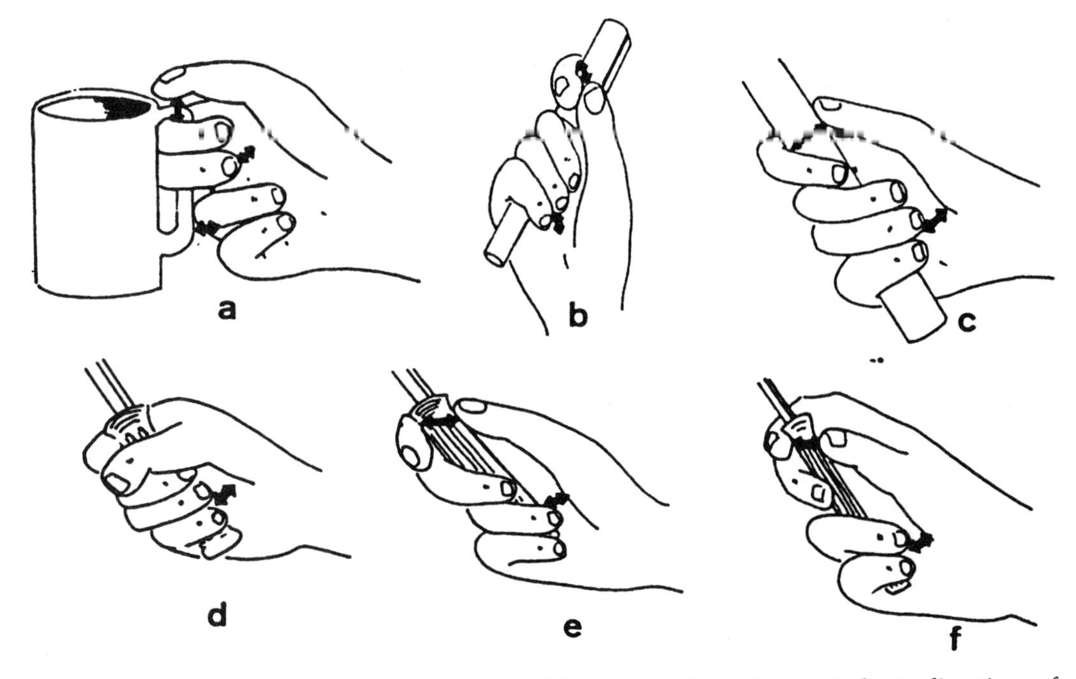

Figure 2–3 Task-specific grasps showing several finger groupings. Arrows indicate directions of force. *Source:* Reprinted with permission from T. Iberall, G. Bingham, M.A. Arbib, Opposition Space as a Structuring Concept for the Analysis of Skilled Hand Movements, *Experimental Brain Research*, Vol. 15, p. 166, © 1986, Springer-Verlag GmbH & Co. KG.

structured, repetitive training,[23] both in terms of their specific problems and as part of their movement rehabilitation,[24] if they are to acquire skill and use these skills in a meaningful way. After a stroke, a person demonstrating impaired physical and social interaction or drowsiness may be judged as cognitively impaired when his or her behavior is the result of depression and/or temporary confusion.

Information and action are interrelated, and the distinction between sensory and motor, as Reed[25] pointed out in the 1980s, has become increasingly blurred. Reed proposed that the dualistic concepts of sensory system and motor system should be abandoned in favor of a concept of an action system. Gordon points out in Chapter 1 that sensation functions in both regulatory and adaptive modes, guides movements during their execution, and corrects movements to improve the next attempt. Gibson[26] also argued that a simple contrast between sense organs and motor organs is probably misleading. Action is object- or goal-directed, related more to environmental circumstances than to receptor stimulation in the strict stimulus-response sense. Emphasis is on the organism as "active" and "information seeking."[27,28] We actively pay attention, ie, engage in perceptual, cognitive, and motor activities when performing a skill rather than merely responding to sensory inputs, and information is selected that is useful to the task in hand. Task-specific training may improve the person's sensory perception concurrently with motor performance because such practice gives the system the opportunity to select and attend to the relevant inputs for the task by providing the need to do so. Although some laboratory investigations have suggested that sensory feedback is not essential to elicit motor output,[29,30] others, most notably Bernstein,[31] have demonstrated that a central command to move is ambiguous without some updated information about changes brought about by the movement itself. Several experiments have shown that even simple reflex movements are modifiable according to context.[32,33]

Many physical therapy techniques have assumed a clear role for the mechanoreceptors in movement initiation and control; yet uncertainty remains about the actual role of muscle and joint receptors in functional movement. Muscle spindles may pick up information about change brought about by movement itself that is used for fine-tuning for precision. The evidence for the traditional view, that joint receptors enable one to sense the position and movement of his or her limbs, is no longer convincing. Several studies on animals[34] and humans[35] suggest that joint receptors may not code limb position or movement but act as limit detectors to signal extremes of range and as pressure detectors.[36] Not enough emphasis has been placed in therapy on verbal feedback (particularly in relation to knowledge of performance) in learning situations and on visual input (in relation to visual cues critical to movement organization and knowledge of results) in providing input critical to regaining motor control and skill.

In the 1980s Newell[37] commented that the field of perception tended to develop in isolation from concepts of action, despite the obvious conceptual links between knowing and doing. Clinical therapy has in large part been based on such a dualistic concept, techniques for improving "sensation" or "perception" being applied to the patient in isolation from self-initiated activities. Sensory stimulation techniques (brisk icing, quick stretch, fast brushing, tapping) are based on localized stimulation of certain receptors of the sensory system to produce a motor response. Passive movement of the patient by the therapist ("handling") is assumed to give the person such kinesthetic information about movement as sequencing of muscle activation and movement speed.[38] However, the relevance to functional action of sensory inputs from passive movement is questionable and no evidence indicates that localized sensory stimulation leads to improvement in functional motor performance.

Several studies[8,39] indicate that specific training of sensory perception (eg, joint position sense, stereognosis) may be necessary, and that such training can result in improved perception of those sensations. Transfer to improved performance of an action may not, however, occur without specific training of that action. The relationship between sensory training and motor performance is an issue that requires testing in the clinic.

Based on available evidence, in training individuals with exercise and practice of specific and meaningful actions involving interaction with objects and the environment, the therapist is training motor output, sensory acuity, and that part of cognitive function that involves paying attention, concentrating, and relating action to intention and to consequences. In setting up rehabilitation so that people also practice without direct supervision, working memory and a sense of responsibility for action are also being trained. The focus is on active engagement by the patient in the rehabilitation process.

The theoretical framework on which we have based our propositions for movement rehabilitation is derived from scientific theories and experimental studies in the general area of movement science, specifically from the fields of biomechanics, motor control, muscle biology, and motor learning. Input for this framework also comes from cognitive psychology and human ecology and from studies of impairments following brain lesions and the adaptations that occur secondarily to the lesion. Increasingly, rehabilitation is based on findings related to the effects of use and experience on the adaptability of the system.[24] In addition, evidence related to outcome following specific exercise and training programs is now emerging, enabling rehabilitation to become increasingly evidence-based.

A theoretical framework is based on certain assumptions about how the system works to control movement, how the organism functions in its environment, and what happens to all this when the system is damaged (see Chapter 1). These assumptions are therefore liable to change as scientific information emerges leading to changes in clinical practice. Developments in knowledge should fuel developments in clinical practice.

THEORETICAL FRAMEWORK FOR MOVEMENT REHABILITATION

In this section we describe briefly under the headings—motor performance, skill acquisition, the rehabilitation environment, the effects of use and experience on recovery processes, and impairments underlying functional disability—some of the material that forms a theoretical framework for movement rehabilitation, including examples of how one might apply this material in the clinic.

Motor Performance

One of the most interesting theoretical perspectives in recent times has developed out of research in various areas showing what, at a certain level, appears rather obvious—that people learn, ie, become more skilled at what they practice. However, to become skilled in performance it is necessary to have sufficient muscle strength (ability to generate and control muscle forces) and muscle length, as well as the ability to coordinate segmental movement to carry out a task successfully. It is clearly evident that patterns of muscle activation and segmental (joint) rotations are task- and context-specific.

The following section comprises a discussion of the muscle in terms of its force-generating capacity and the specificity of muscle activation patterns occurring during muscle strengthening. The second issue discussed is motor performance, which is exemplified by the performance characteristics of balancing, reaching, and manipulation, derived from biomechanical studies.

Muscle Activity: The Generation and Control of Force

A muscle or muscle group must be capable of generating sufficient force to bring about a movement whether intended or in response to an unexpected perturbation. Force has to be timed and graded appropriately to enable control of synergic muscle action. Muscles also need the endurance to generate force over a relatively long period (eg, walking up several flights of stairs). Muscles must be capable of building up force fast enough to meet the demands of the environment (eg, crossing the street at traffic lights) and of the task (eg, using a keyboard with accuracy). Muscles also need the capacity to lengthen according to action demands. Normally we have the strength, muscle extensibility, and cardiovascular fitness needed for our daily lives. However, when we want to do an activity that requires additional muscle length, muscle strength, and fitness we undergo special training. The link between our own muscular capacity, endurance, and functional performance is obvious to us every day.

Following an acute lesion affecting the central nervous system (CNS), patients experience muscle paralysis and weakness due to disruption of the motoneuronal input activating muscles. In addition, it may not take long for the individual to become deconditioned with generalized muscle weakness, muscle atrophy, and decreased cardiovascular fitness that are known to be associated with imposed inactivity.[40,41] Given that stroke is principally seen in older individuals, it is also likely that many individuals were inactive and physically unfit before their stroke.

Several recent studies have shown beneficial effects of muscle strengthening in neurorehabilitation. Significant relationships have been found between measures of muscle strength and measures of motor performance and functional activity in adults following stroke, in children with cerebral palsy, and in elderly persons following musculoskeletal lesions. For example, correlations have been found between the following:

- Grip strength and Motor Club assessment scores[42]
- Wrist extension strength and upper limb function[43]
- Lower limb muscle strength and Fugl-Meyer motor scores,[44] gait,[17,45–47] stair climbing,[48] balance in standing[49,50]

Significant increases in muscle strength and functional ability have been reported in elderly persons following strength training.[16,51,52] Physical conditioning programs have shown positive effects in older individuals and following brain lesion.[53–56]

In individuals with brain lesions, increased strength may be a result of increased "efficiency" of motoneuron recruitment.[57] Strength training may therefore have a neural retraining effect.[58] It is also likely that increased strength may result from increased mechanical efficiency of the muscles themselves.

Interestingly, the positive effects of strength training do not necessarily transfer from one action to another or from one context to another. An early experiment by Rasch and Morehouse[59] that described the results of a study in which the effect of exercise to increase strength of elbow flexors, performed in standing, found that able-bodied persons demonstrated an increase in strength when they were measured in standing but a lesser increase when they were measured in supine lying. That is, there appeared to be little transfer of training effect from one position to another. The authors pointed out that this result may have reflected the learning component of exercise. It is also evident on subsequent reflection that the muscular coordination to stabilize the body to allow a forceful contraction of elbow flexors in standing would differ from the coordination required in supine. Some muscles will be performing postural adjustments, and others will be stabilizing. For example, postural muscle activation patterns in standing would have involved lower limb muscles, the contraction of which subjects would have learned to time before as well as during the weight-lifting exercise to control the total body perturbation brought about by the exercise. Therefore, it is not so much that the elbow flexors in isolation are being strengthened in this experiment but that a number of muscles are being trained in a manner that is specific to the task and the position in which the task is being performed.

This early work illustrated the posture-specificity of muscle action. Other work demonstrates angle, velocity, and muscle contraction type specificity.[60–65]

Some evidence indicates that transfer of muscle strength from open chain exercise to closed chain actions is limited. Everyday actions are typically carried out by concentric (shortening) and eccentric (lengthening) modes of muscle contraction. Most of the actions in which the lower limbs are involved are closed-chain (ie, the body mass is raised and lowered over a fixed base of support—the feet) in their mode of action (eg, standing up and sitting down, stair ascent and descent, propulsive stance phase of walking). Rutherford and colleagues[62] found that able-bodied persons who were trained to lift a load with knee extensors through range (open-chain exercise) were able to increase the load lifted by 200% over 12 weeks. Maximum isometric force, however, increased by only 11%. Further, when maximum power output generated during isokinetic cycling (closed-chain action) was measured after the strength training, no increase was found either at slow or fast speeds, even though the knee extensors are major contributors to pedal movement.

Similarly, studies in which concentric and eccentric actions are compared show some different effects between the two modes of action. For example, eccentric action produces greater loading of elastic components of muscle;[66] lesser levels of muscle activation for greater levels of force are produced in eccentric work compared with concentric;[63] eccentric contractions appear to involve lower motor unit discharge rates than concentric contractions.[67]

Nevertheless, it appears that when muscles are very weak (eg, after an upper motoneuron lesion), any exercise designed to increase strength will be effective initially. Once a certain threshold has been reached, however, strengthening exercise needs to be task-specific because it is skill in performance that is most relevant at this stage.[68] In the clinic, this information would infer that, when muscles are very weak, simple exercises involving open- and closed-chain exercises, eccentric, concentric, isometric, and through-range modes of action, using, for example, electrical stimulation or machine-assisted movement, are likely to be effective at helping the patient to activate a muscle group and increase the muscle's ability to generate force. However, once a certain threshold is reached, strengthening exercises should incorporate similar performance characteristics to the action to be learned; eg, stepping exercises, repetitive sit-to-stands with incremental lowering of seat height, practice of the action in the appropriate environments and with appropriate goals to learn to apply the muscles' improved force-generating capacity.

Critical components. Biomechanical studies examining the structure and timing of segmental movement in common actions such as sit-to-stand, walking, and reaching suggest that there are some components of an action that are free to vary and others that are invariant if the task is to be performed effectively in different contexts and environments. In earlier work,[69] we expressed the view that those kinematic components which are critical to performance of a task provide guidelines for training in clinical practice. Critical kinematic components are identified by recording movements of body segments on videotape using anatomical landmarks and motion analysis systems as subjects perform a specific action. Computer software is used to analyze the recordings and calculate angular and linear displacements, velocities, and accelerations.

Critical components as we have defined them are observable joint angular displacements and linear paths of body parts (shoulder path in STS, foot path in swing phase of gait, hand path in reaching to an object) developed from biomechanical studies and replicated under different conditions. These kinematic variables are reflective of the underlying momentum and force characteristics that are not observable without complex instrumentation. Interestingly, they tend to be consistent across subjects and conditions, although there may be variability at the level of force production and muscle action. It is the consistency of the kinematic components, suggesting that they are critical to effective performance, which makes them appropriate guidelines for clinical practice. These guidelines can be used in the clinic not only for analysis of performance but also as a basis for training.

Whereas only limited interpretation of motor deficits can be made based on kinematic data alone, angular displacements and linear paths do form a repeatable, invariant pattern in skilled actions and, therefore, enable comparison to be made at that level. Observational analysis without such a data-based formalized structure tends to be arbitrary. As Saunders and colleagues[70] pointed out in a classic article in which they investigated what they called the biomechanical determinants of gait:

> It is our expectation that, by an appreciation of these fundamental determinants, the orthopedic surgeon [*read physical therapist*] will be able to analyse disorders of locomotion with greater precision and to apply corrective measures with a fuller understanding of the interrelationships which exist between the various segments of the locomotor mechanism. (p 543)*

As an example, we think there is sufficient evidence from biomechanical studies of sit-to-stand under different task and environmental conditions to suggest that the critical components of sit-to-stand (Figure 2–4) are:

- initial foot placement backward (approximately 10 cm from a perpendicular line dropped from the knee joint)
- flexion at the hips with trunk extended; dorsiflexion at the ankles
- a sequence of lower limb extension (knee, hip, ankle)

The second component is the major factor producing the horizontal momentum of the body mass that is transferred into vertical momentum produced by the lower limb extension. The last two components are associated with a rapid increase in vertical ground reaction force brought about by extensor forces produced at the hip, knee, and ankle.

**Source:* Reprinted with permission from J.B. Saunders, V.T. Inman, and H.D. Eberhart, *The Major Determinants in Normal and Pathological Gait,* Vol. 35A, pp. 543–558, © 1953, Journal of Bone and Joint Surgery.

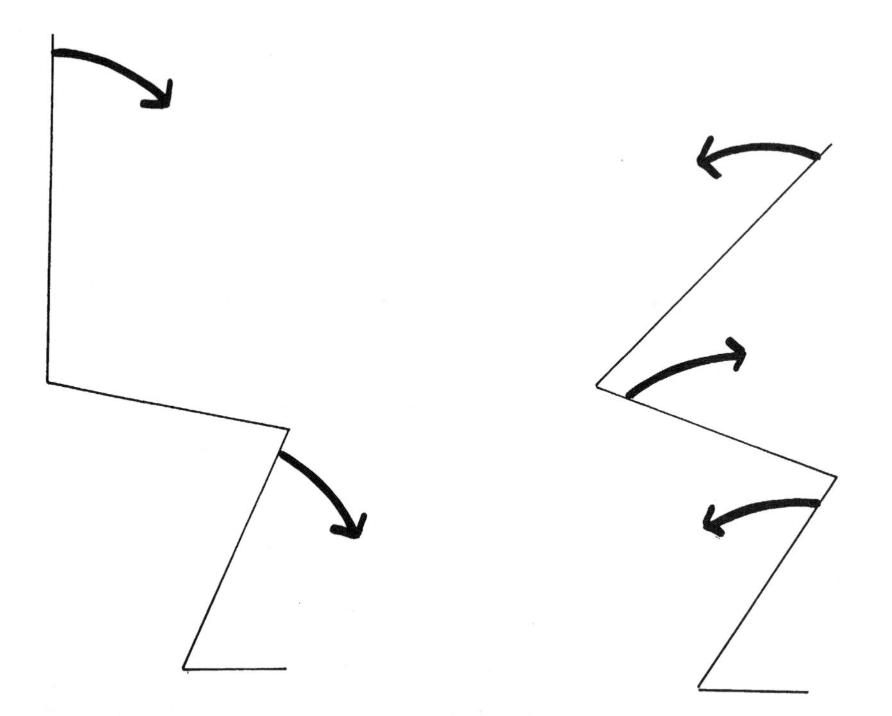

Figure 2–4 Stick figures illustrating the direction of segmental rotation during standing up. L: The body mass is moved forward by flexion of the trunk at the hips with clockwise rotation of the shank on the fixed foot. R: Vertical movement of the body mass involves extension at the hip, knee, and ankle. Hip extension occurs through both anticlockwise rotation of the trunk and clockwise rotation of the thigh, and extension of the knee by clockwise rotation of the thigh and anticlockwise rotation of the shank on the foot. In standing up the trunk is extended throughout the action, behaving as a single "virtual" segment as it rotates forward and then backward at the hips. Although some movement is likely to occur at spinal joints, biomechanical analysis indicates that this movement is of relatively small excursion.

If standing up is to be carried out flexibly and easily in different environments and for different goals, these components must be present. It is possible to stand up with the feet further forward if you are young and fit, but there is a cost in terms of the muscular effort required.[71] Standing up from a higher seat requires less extensor muscle force than from a lower seat.[72] Movement of the trunk forward by flexing at the hips is the major contributor to the horizontal momentum that transports the body mass forward over the new base of support.[73] A greater overall magnitude of extensor force is produced at hips, knees, and ankles as the body mass is raised vertically when active trunk movement is absent[74] and when arm movement is constrained.[75] The implications for clinical practice are clear. Placement of feet backward, modification of seat height, and starting the action from an erect position and leaving the arms free to move instead of holding them over the chest are guidelines for training standing up.

Balance

Balance is the process by which the body's equilibrium is controlled for a given purpose.[76] It is the ability to control the body mass or center of gravity relative to the base of support. Linked segment dynamics play an important role in controlling balance. The adjustments we make to preserve balance are flexible and varied because of the potential for dynamic interactions afforded by the body's segmental linkage. A balanced body mass is achieved by postural adjustments, ie, muscle activation patterns and segmental movements enable us to control the segmental linkage in relation to the base of support. In daily life balance can be disturbed in three ways: by an external force applied to the body; via support surface movement; and by internal forces applied during a self-initiated movement.[24]

Postural adjustments are specific to task and context. Many studies have shown that, in standing, anticipatory postural adjustments before voluntary movement and ongoing adjustments appear to depend on initial conditions,[77,78] the context of the action,[79-82] speed and amplitude of movement,[83] and the amount of support provided.[77,84,85] Of considerable clinical relevance is the fact that conditions of support change the locus of balance control and that very slow movements are not as perturbing as faster movements.[83]

One's ability to balance emerges from a complex interaction between sensory and muscular systems integrated and modified within the CNS in response to changing internal and environmental conditions. The relative roles of different sensory inputs remain controversial but they are probably coordinated in a task-appropriate manner. The vestibular system provides information about the position of the head in relation to gravity and to motion. The proprioceptive system (muscle, joint, and cutaneous receptors) provides information about the state of the effector apparatus, relative position in space, and such environmental factors as surface conditions. For example, plantar cutaneous afferents play a significant role in balance control in standing[86] and it is likely to be so in sitting also. The visual system provides information about the orientation and movement of the body relative to environmental features.

Ability to maintain equilibrium while relatively motionless and during movement involves: 1) generation of muscle activity to support the body mass against gravity and prevent collapse; 2) control of body segments in relation to each other; and 3) control of body alignment in relation to the environment to maintain the body's center of gravity within the limits of stability.[24] The integrity of musculoskeletal components (eg, extensibility of soft tissue, active and passive properties of muscle) and the ability to generate sufficient muscle force (muscle strength) are critical to one's ability to balance.

Standing is an intrinsically unstable position[87] in which one maintains balance (ie, does not fall over) by complex interactions of muscle activations that ensure that the destabilizing effects of volitional movement (reaching to pick up an object, taking a deep breath) and of external events (moving walkway, motion of a bus) are controlled for. That is to say, one does not fall over unless some unforeseeable or extreme event occurs. Sitting is, however, a relatively more stable position, particularly when the feet are on the floor rather than hanging free.[88] With the feet on the floor, the lower limbs provide an active as well as a passive base of support and lower limb muscle activity is therefore necessary to one's ability to balance in sitting under different task conditions, eg, reaching to objects beyond arm's length.[89] In reaching from a small base of thigh support, the legs are more critical to balance than when thigh support is greater.[89]

The limit of stability is reached sooner when the body mass is moved laterally than when it is moved forward because reducing the area of the base of support decreases the region of stability. There is a perimeter beyond which, in standing, one cannot move the body mass without taking a step (ie, making a new base of support) or overbalancing and falling. The region in which one can balance while moving is called the "region of reversibility,"[90] the area before one reaches the "limit of stability."[91] One's "perceived" limit of stability may be different if one perceives a threat to his or her balance by misinterpreting visual inputs and by fear or apprehension.

In the area of postural adjustments, several studies have shown that the order in which muscles turn depends on the task being performed and its environmental context.[77,92,93] For example, when subjects stood on a movable platform holding a stationary lever and the platform was unexpectedly perturbed, the order of muscle onset was arm muscles before leg muscles. However, when the freely standing subjects were asked to pull on the hand lever, leg muscles were activated before arm muscles, the order of muscle onset being soleus, biceps femoris, and biceps brachii.[77,93] These results indicate that the muscles that are turned on for postural support or stability are those that can best provide it, and this depends on the task being performed and on the possibilities for support (Figure 2–5). In other words, even in standing postural muscles are not necessarily leg muscles because any body part can potentially serve as a base of support or point of stability. The findings suggest that interpretation of information from a variety of inputs and from output motor commands is to a large extent planned according to perceived support conditions, ie, they are based on perceptions of how the environment is structured and the intention of the action.

In a study of the effects of visual information on muscle onset latency, Dietz and Noth[94] had their subjects fall forward on to a board that could be adjusted at various angles to the perpendicular. The experimenters monitored

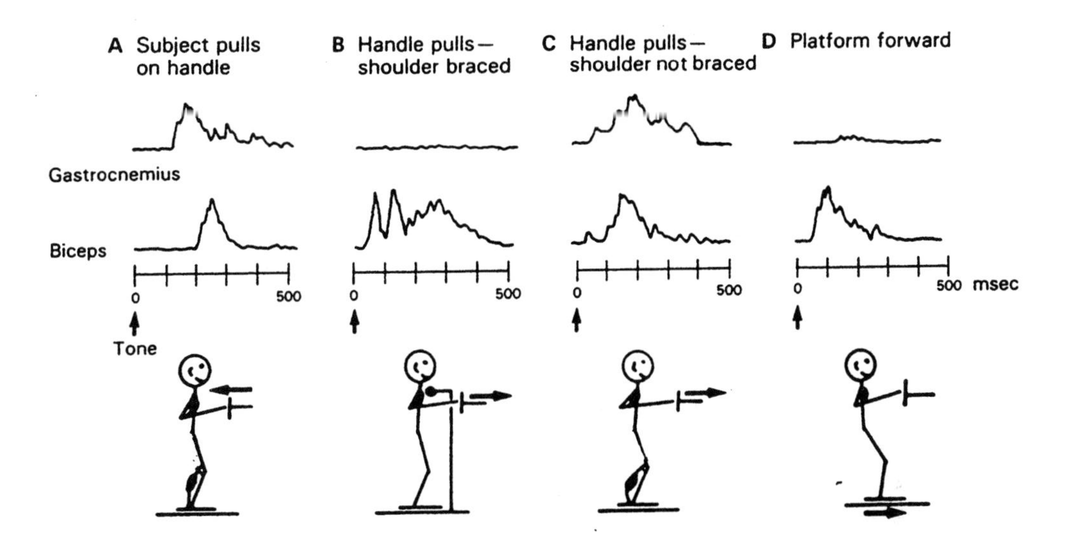

Figure 2–5 The relative onsets of arm (biceps brachii) and leg (gastrocnemius) muscles under different task conditions in standing: (A) when subject is asked to pull on a handle; (B) when the handle is unexpectedly pulled but subject is leaning against a chest support; (C) when the handle is unexpectedly pulled when the subject is free standing; and (D) when the platform is unexpectedly moved forward. *Source:* Reprinted with permission from J.H. Carr and R.B. Shepherd, *Neurological Physiotherapy Optimizing Motor Performance,* p. 186, © 1999, Butterworth Heinemann.

the triceps brachii muscles electromyographically. No matter at what angle the board was placed, ie, no matter how far the subjects had to fall, electromyography (EMG) activity was evident a constant amount of time before impact. When the subjects were blindfolded, however, the triceps turned on at a constant duration after the instruction to move. Visual information about time to contact when falling forward appeared to be the critical input by which the system controlled the effector apparatus.

Vision and its critical role in balancing has been explored by Lee and his colleagues,[27,95] who proposed that vision could be considered "exproprioceptive" because from vision one gains information about his or her position relative to the environment, a theory put forward by Gibson.[26] The experiments of Lee and others have shown that visual information can override proprioceptive (kinesthetic) information from muscles and joints under certain circumstances. In one experiment,[96] 14 subjects standing in a darkened room were monitored to see the effect on their body alignment of the movement of an illuminated rod. The experimenters found that as the rod moved so did the subjects, suggesting that in this situation visual information was overriding both kinesthetic and vestibular information about position in space.

The information presented above is very compelling in terms of developing training strategies for the clinic. In developing training strategies it is also necessary to correctly analyze the motor dysfunction and consider the nature of the impairments, eg, some current therapy aims at improving trunk control and facilitating trunk movement [38,97,98] This therapy appears to be based on two assumptions: 1) that poor trunk muscle activation is a major impairment following stroke and that poor trunk control underlies the patient's balance difficulties; and 2) that improving trunk control will transfer into improvement in function. However, no experimental or theoretical support exists for these assumptions. We agree with Dickstein and colleagues[99] that the rationale for the current emphasis in physical therapy on restoration of the function of trunk and girdle muscles is puzzling because it is the distal muscles rather than the proximal axial muscles that are affected principally in hemiparesis.[100] Axial muscles receive contralateral as well as ipsilateral descending inputs and a recent study provided no evidence that trunk muscle activation was impaired in the hemiparetic subjects studied.[99] Further, it is likely that weakness (or paralysis) and poor control of lower limb muscles underlie difficulties balancing in both sitting and in standing. It is these muscles that control the movement of the trunk over the base of support (thighs and feet in sitting, feet in standing). The decreased trunk movement seen as a person reaches out in sitting (see Figure 2–7), is likely in many patients to reflect a reluctance to move the body mass too far toward the periphery of the base of support because of insecurity that results from difficulty stabilizing the lower limbs.

The experimental evidence is that postural adjustments and focal limb movements (posture and movement) are interrelated, interdependent, and specific to task and context.[101] It is for this reason that we proposed that therapists move away from the use of therapeutic procedures to improve trunk control and balance in isolation from real-life activities, an approach that appears to be based on the dualistic assumption of a postural system separate from a movement system. However, training balance as part of training critical everyday actions is based on the assumption of a dynamic system controlling all aspects of a motor task according to the requirements of that task.[102]

We proposed more than a decade ago[69] that a person with a brain lesion affecting motor areas that has difficulty activating or controlling the necessary muscular synergies to move effectively in sitting or standing, needs to practice in these positions. Where balance is very poor and the individual is anxious, simple tasks that cause only small displacements of body mass can be practiced initially. Practice progresses to a variety of tasks (Figure 2–6; Figure 2–7; and see Figure 2–29), particularly reaching actions, that require increased excursions of the body mass, increased speed, and increased complexity (eg, different objects, different tasks). The objective is to train the individual to increase the perimeter in which the body mass can be confidently, effectively, and safely moved. If a person's standing is impossible because of lower limb weakness that causes the limb to collapse, a light wraparound splint holds the knee in extension (preventing collapse at the knee and taking away the need for adaptive knee hyperextension), reassures the patient, and enables practice of actions in standing.

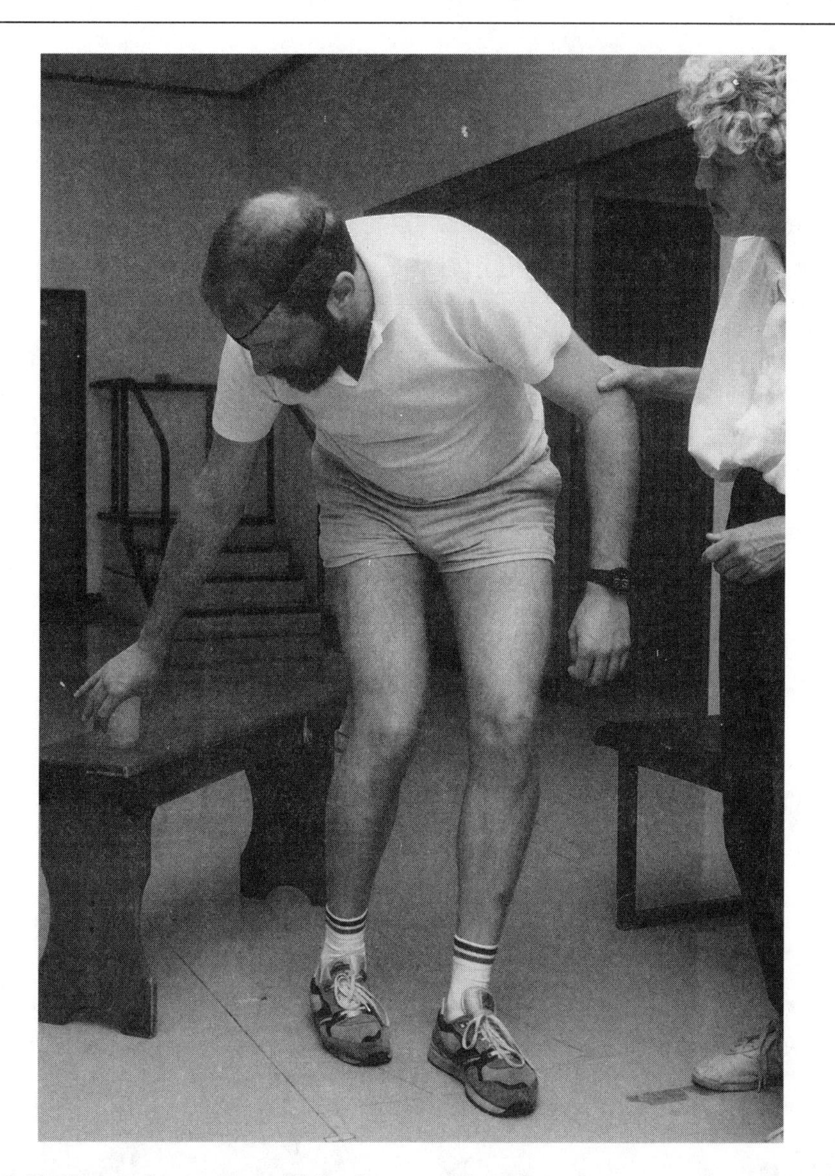

Figure 2–6 Flexion and extension of hips, knees, and ankles over a narrow base to reach to pick up an object. The exercise provides practice of eccentric and concentric lower limb extensor muscle activity during a difficult balancing task. *Source:* Reprinted with permission from L. Ada and C. Canning, *Key Issues in Neurological Physiotherapy*, © 1990.

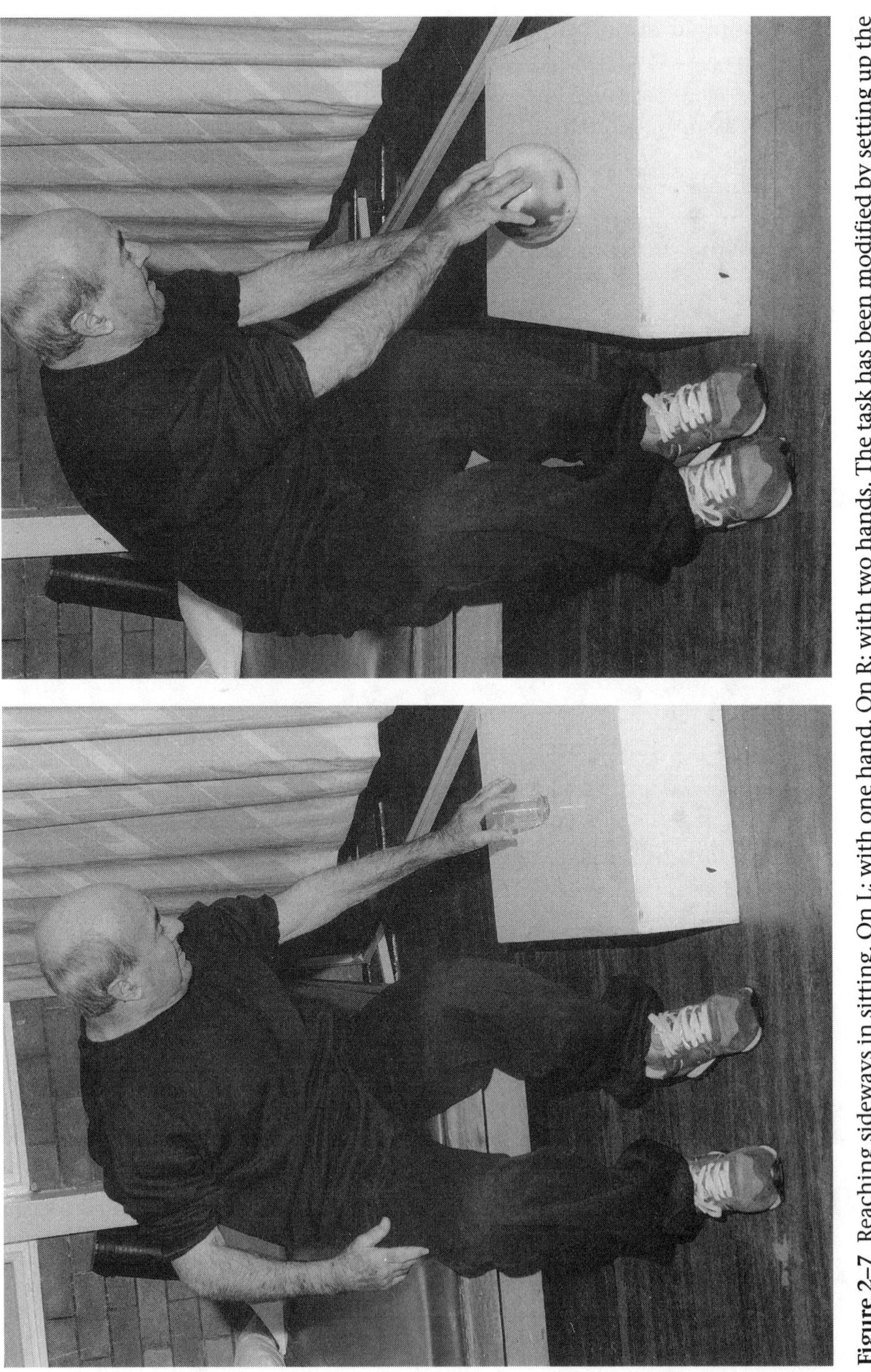

Figure 2–7 Reaching sideways in sitting. On L: with one hand. On R: with two hands. The task has been modified by setting up the environment to require less lateral movement of the body mass because he cannot reach down to the floor. Weakness of his left leg muscles makes it difficult to stabilize the leg. Reaching to the left gives him practice of loading the left leg and using it to shift his body mass back to the upright sitting position.

Practice of reaching tasks requires that the system controls not only the arm movement but also the postural adjustments that are interrelated with the arm movement. It is probable that it is only by practicing in this way that the system has the opportunity to re-establish these relationships. Early training and practice can take place in a suspended body harness, with the individual having confidence that they will not fall[12] (Figure 2–8). The use of a harness may be essential for patients who are unable to stand without being supported by the therapist. Practice of simple actions can take place within the security of the harness during group work and circuit training. The patient can be more active if the therapist does not have to provide support manually. A common tendency in rehabilitation is to "protect" patients, to make them physically and emotionally secure. Standing may not even be practiced if the therapist sees any risk of a person falling over. If standing is attempted, fear of the patient falling over or of the leg collapsing may lead the therapist either to support the person or to encourage them to hold on to a rail for support. The person is unlikely to have the opportunity to improve the ability to balance in either situation.

Muscle weakness and paralysis as well as the fear of falling that occurs as a result have serious effects on a person's ability to move about in standing and in sitting. Strengthening exercises for the lower limbs (eg, Figures 2–25 and 2–26) are critical to improving the ability to support the body mass and to balance in standing. The relationship between muscle weakness and poor balance seems not to be well understood in clinical practice. Mounting evidence indicates, however, that strengthening exercises for lower limbs result in improvement in walking (and, by inference, balance while walking) and fewer falls.[7,16,103]

From the available evidence, the following appear to be the critical factors in optimizing the ability to balance the body mass, ie, to move without falling over:

- An early start at practice of balancing in erect positions (sitting and standing).
- Intensive practice of simple movements that require only a small excursion of the center of body mass, such as moving the head (to view parts of the room and report back on details observed) and reaching for objects. As the person gains more control, larger excursions are demanded by tasks in which the person reaches beyond arm's length. More complex movements are practiced such as bending down to pick up objects from the floor. Practice of walking and of actions such as standing up and sitting down enlarge the person's repertoire of balanced actions.
- Harness support should be available to enable practice by patients who cannot otherwise stand independently (Figure 2–8). Similarly, walking can be practiced with supportive harness and on a treadmill (Figure 2–9).
- A computerized force plate system provides motivating feedback,

Figure 2–8 A harness provides safety during practice of reaching sideways to place the ball on the table. In reaching to the right he is practicing using the left leg to move his body mass over to the right.

---

enabling the patient to practice enjoyably if the body mass movements provide critical information or are linked to a game.

- Exercises to strengthen lower limb extensor muscles (particularly hamstrings, quadriceps, and calf muscles) and improve control of the lower limbs in actions with the feet on the floor, in particular, raising and

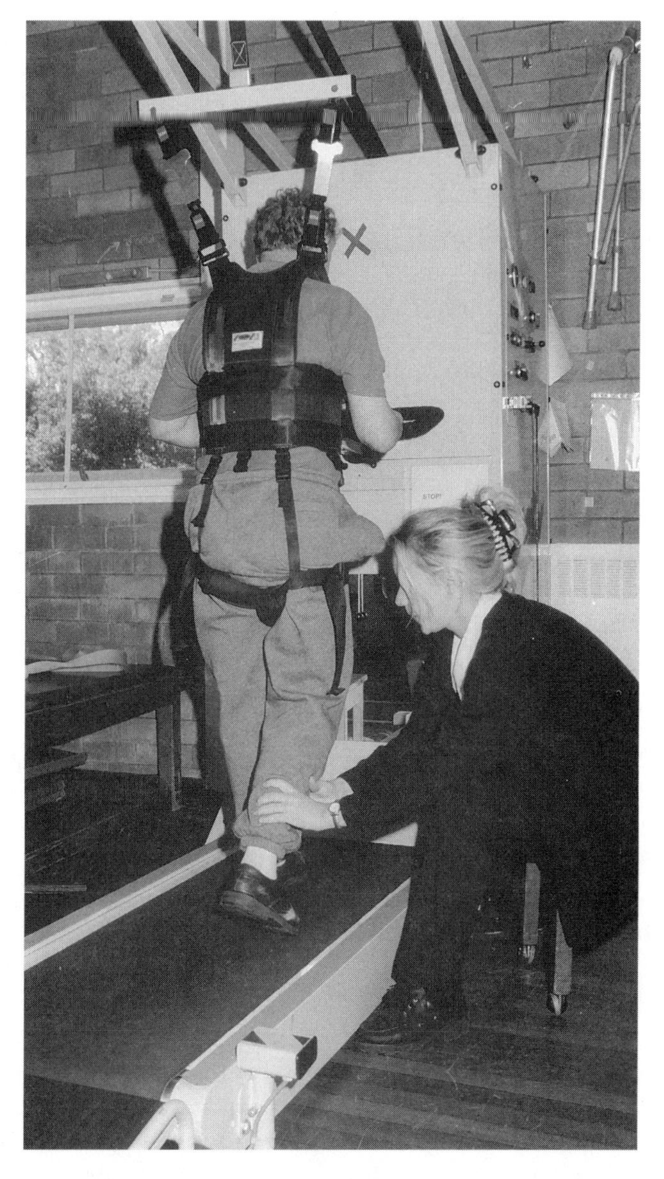

Figure 2–9 Treadmill walking with supportive harness. The therapist assists with toe clearance and step length during swing phase of the hemiparetic leg.

lowering the heels, lateral and forward stepping exercises (Figures 2–24 and 2–25).

- Practice of activities that increase the person's ability to function flexibly in changing environments, catching or kicking a ball, hitting a ball with a bat or racquet, and standing on a moving support surface.

- A stretching program is instituted if muscles, particularly calf muscles, are short or at risk of becoming short.
- Activities should challenge people to the limits of their stability for improvement to take place. Use of arms for support should be discouraged.

Reaching and Manipulation

The hand is the major interface with the world external to ourselves. It is transported to the scene of the action by reaching, an action that involves all segments of the upper limb. One can extend his or her reach beyond arm's length by movement of the upper body and, in standing, by taking a step or walking. It is evident that the arm and hand function as a single coordinated unit,[19,104–106] and that—where relevant—upper body and total body movement become part of that unit.[107,108] Grasp and hand orientation during reaching reflects what is to be done with the object[109,110] and the nature of the object, eg, its fragility, slipperiness, and shape.[111–114]

As Gentile points out in Chapter 3, the arms are also yoked into the postural system. Depending on support conditions, manipulation requires varying degrees of postural adjustment to balance the body mass. When balance is at risk, the arms play a stabilizing and supportive role, and if balance is lost the hands may be used to form a new base of support. Postural adjustments before and during arm movement have been reported in standing and recently in sitting, in both fast-[115] and self-paced[107] movements. Although reaching to an object in sitting can be achieved theoretically by a variety of movement combinations, it appears that the most parsimonious solution is used. Two recent studies have shown a consistency of movement pattern across subjects when they reach forward in sitting to pick up an object beyond arm's length.[107,116] Reaching to an object *within arm's length* involves the upper limb, although the upper body (trunk), if it is not supported, also moves to a small extent. Reaching *beyond arm's length,* however, requires movement of the trunk (at the hips) to lengthen reaching distance.

Many clinically relevant studies of reaching have been conducted by experimenters whose major objective is to examine various hypotheses regarding mechanisms of motor control. Although major questions remain unanswered, these studies illustrate the complexity of upper limb movement and provide support for specificity of training. Whether the central command is formulated in terms of trajectories of body parts (eg, the hand path in reaching) or in terms of joint angular displacements, both trajectory and angular displacement are guided externally by the task requirements (eg, position of glass to be picked up, shape of glass, its weight, and what is to be done with it).[20,109–111]

The path of the hand as a person reaches to pick up an object is shown in Figure 2–1. Reaching to grasp an object can be divided into two components: a transportation phase and a slower manipulation phase that involves final adjustment to grasp aperture made just before grasp.[105,106] The hand starts to open at the start of the reaching action.[19]

Most commonly performed actions are carried out with both hands and the interactions between the two hands and the object to be manipulated are complex[117,118] (Figure 2–10). One experiment has shown that when a person holds a ball and lets it drop into a cup held by the other hand, the grip force of the cup hand increases in anticipation of the ball's impact.[119]

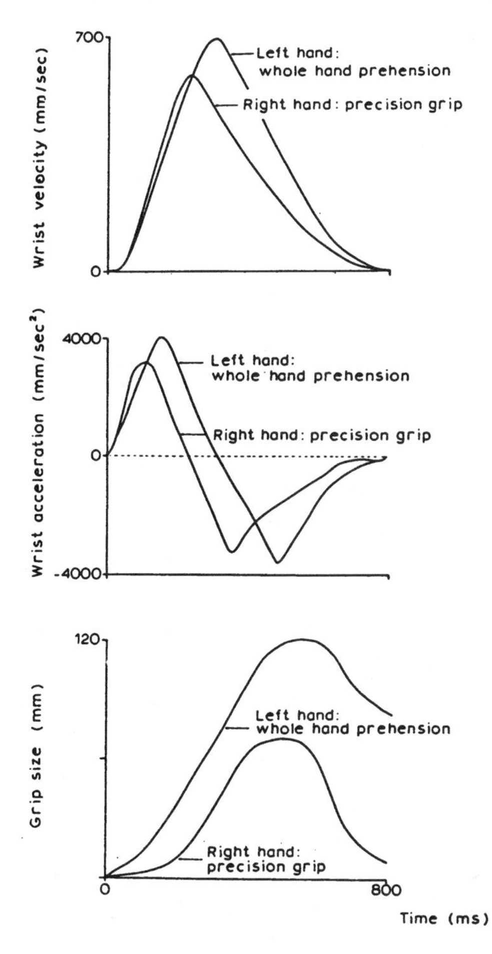

Figure 2–10 A single trial of a bilateral task—reaching to grasp a can (left hand) and pull the tab (right hand). (Top) wrist velocity, (middle) wrist acceleration, (lower) grip size. Both hands reach the can together; however, peak velocity, peak acceleration, and deceleration occur earlier for the precision grasp hand than for the whole hand grasp. *Source:* Reprinted from *Behavioral Brain Research*, Vol. 56, U. Castiello, K.M.B. Bennett, and G.E. Stelmach, The Bilateral Reach to Grasp Movement, p. 48, © 1993, with permission from Elsevier Science.

The anatomical structure of the hand and the cortico-motoneuronal connections to hand muscles[120,121] allow a variety of possible movement combinations. Manipulation is affected, as is the reaching action, by the object to be manipulated and the intention or goal. When the fingers move to manipulate an object, the dominant pattern is one of fractionation of movement. However, when force is required for gripping, a pattern of coordination emerges and the number of muscles involved depends on the degree of force exerted. That is, the articulated anatomy of the hand and the motoneuronal control of the muscles enable the various components of the hand to function relatively independently either with movement of individual fingers or with fingers moving as a total unit. Details of specific interactions involved in different tasks are given in many articles[121–124] and as an overview.[24]

Work by several authors suggests that the control of hand movement is simplified by the control system. The hand is "shaped" according to the manipulation required.[121] Groups of fingers act as a simple functional unit in what has been called a "virtual finger"[125,126] (see Figure 2–3). This concept of neural and mechanical simplification is useful in clinical practice. It is possible that training control of functional units used in many different tasks might transfer into improved performance of tasks made up of those units. This information enables the formulation of critical actions to train that will have maximum chance of transferring into improved performance of a number of related tasks.

Although reaching for an object appears to be carried out under visual control, once the hand contacts the object, sensory inputs come from tactile and pressure sensors in the hand. These inputs enable us to localize stimuli and appreciate fine detail. Tactile information enables us to monitor the weight of an object and its frictional characteristics in relation to slip and to adjust motor output through the "vertical lifting force" and the "slip-triggered grip force."[127–129] Even moving the hand while holding an object produces inertial forces that are compensated for by changes in grip force.[130]

In clinical practice, an understanding of the control and biomechanics of reaching for and manipulating objects provides a framework for developing a training program. Major constraints to training are the muscle paralysis and extreme weakness that are common in patients having upper limb involvement following stroke. Further constraints come from the nature of postlesion motoneuronal control of the hand, the speed with which soft tissues adapt to disuse and prolonged positioning, and the natural tendency for patients to focus on their unaffected hand for their daily activities.

Secondary musculoskeletal complications, resulting in shoulder stiffness and pain and glenohumeral subluxation, are common sequelae following stroke. Shoulder pain is known to be a major cause of poor recovery, interfering with rehabilitation and causing great distress.[131,132] The principal causes are the weakness and paralysis of muscles and the immobility of the limb, with disuse-provoked adaptive changes in muscle and other soft tissue. The resulting adhesive capsulitis may underlie pain and stiffness in many patients. Every effort should be made to prevent or at least minimize the development of secondary adaptive changes. Some evidence indicates that persistent maintenance of glenohumeral internal rotation and adduction, with adaptive shortening of muscles that link the scapula to the humerus, in particular internal rotator muscles, is a major causative factor in pain development. Preventive passive stretching by periods of positioning (Figure 2–11), active movement with assistance if necessary (Figure 2–12), and modified training of reaching (Figure 2–13) should be a mandatory part of rehabilitation, with periods of the day set aside to achieve the goal of a pain-free shoulder.[24]

Results on the whole suggest that recovery of upper limb function is minimal in many patients.[133,134] However, it is also evident from modern physical therapy texts[38,135] that active upper limb rehabilitation is neglected. It is inter-

Figure 2–11 Positioning procedure to prevent changes in length of soft tissues around the glenohumeral joint, in particular adductor and internal rotator muscles that are particularly vulnerable. A sandbag can be used to keep the arm in position if necessary.

Figure 2–12 Arm cranking. A method of gently mobilizing the affected elbow and shoulder and promoting muscle activity. The hemiparetic hand can be bandaged on if necessary.

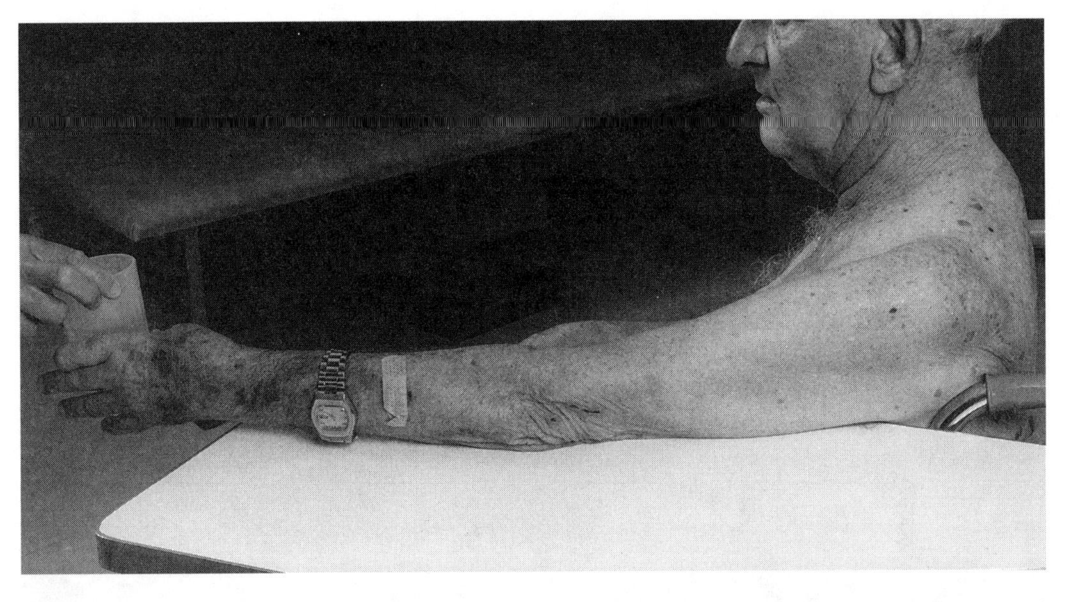

Figure 2–13 Reaching with the arm on a table reduces force requirements of muscles around the glenohumeral joint, particularly deltoid muscle, and enables practice of active reaching to an object.

esting that studies showing more promising results[136–138] were conducted with patients whose rehabilitation was active and task-related.

It is evident that only a small percentage of the day is spent in intervention relevant to the upper limb. One audit in a rehabilitation center reported 10 minutes per day.[139] Much of the patient's day may be spent with the hemiparetic limb in a sling or in propelling a one-arm drive wheelchair with the unaffected upper limb. Such experiences ensure that patients become increasingly proficient at using the unaffected limb.

From the available evidence, the following appear to be the critical factors in optimizing outcome in terms of upper limb function:

- An early start to exercises and other techniques to stimulate muscle activity. If active exercise is not possible because of paralysis, electrical stimulation applied to groups of muscles around glenohumeral joint and hand (eg, wrist, finger and thumb extensors, abductors and opponens) may enable the patient to get the idea of muscle contraction and to activate apparently paralyzed muscles (Figure 2–14).[140–143] Electrical stimulation will also act to preserve the muscle's contractility and extensibility. EMG feedback can be effective in enhancing a person's ability to contract a muscle[144] (Figure 2–15). Computer-generated feedback (with computer

Figure 2–14 Electrical stimulation of paretic wrist extensors to preserve the integrity of the muscles and promote muscle activity. The ruler provides a goal for active wrist extension.

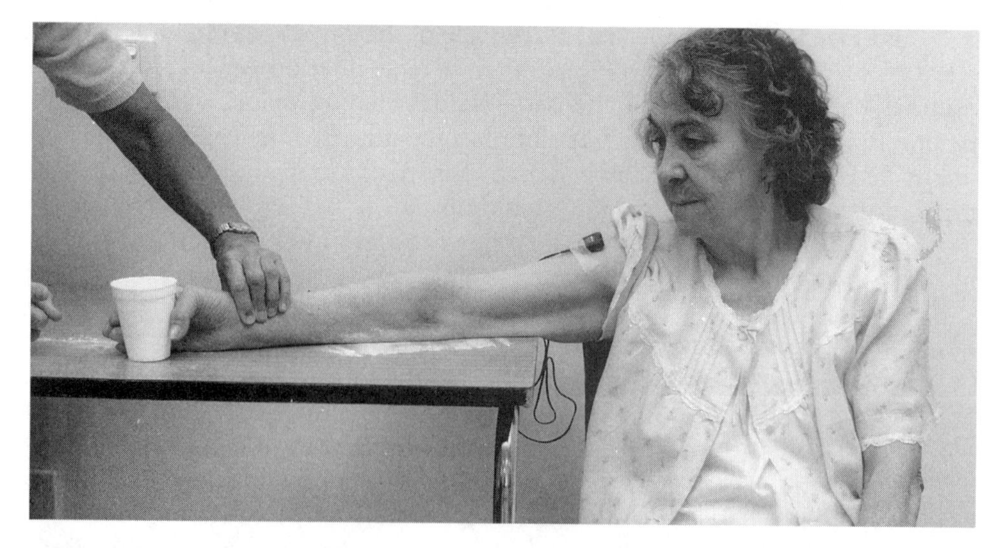

Figure 2–15 Using an EMG device to monitor activity in deltoid muscle during attempts at reaching forward. An auditory signal gives information about presence or absence of muscle activation.

games) appears to have considerable potential for driving activation in apparently paralyzed muscles[145] and for improving muscle force generation (Figure 2–16). Devices can be set up that enable practice of exercises directed at critical muscle groups (Figure 2–17).

- An early start to preserving muscle length and preventing increased muscle stiffness and preventing injury to the unprotected glenohumeral joint, while muscle paralysis is present. Adaptive muscle shortening (particularly of glenohumeral internal rotators), together with muscle

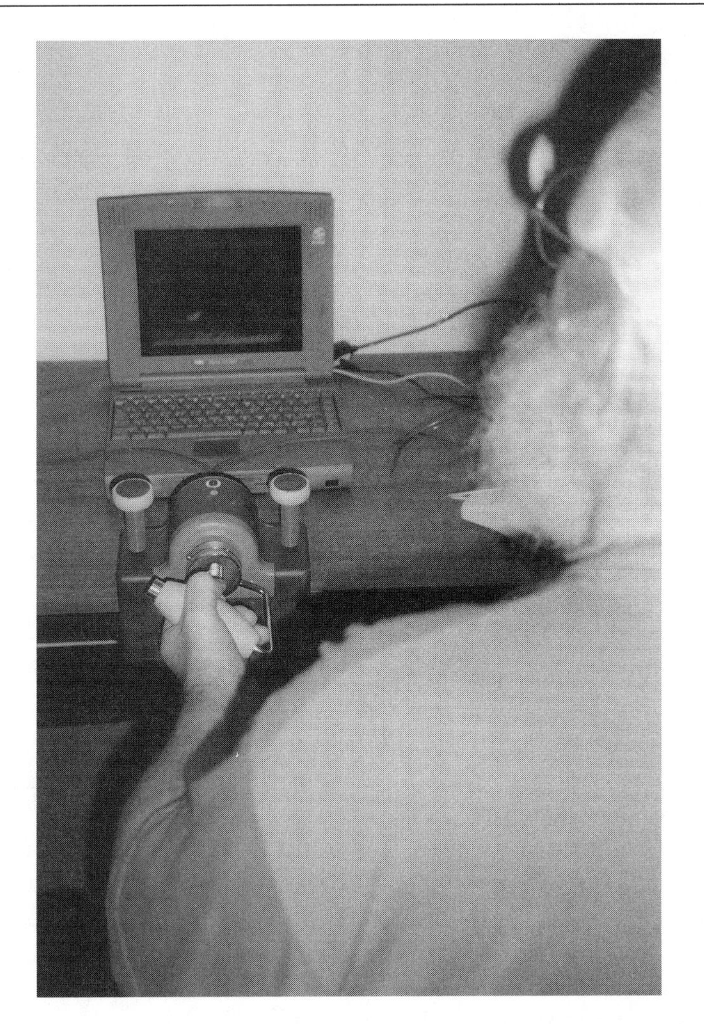

Figure 2–16 Independent practice of supination and pronation on the Upper Limb Exerciser (Biometrics Ltd, PO Box 340, Ladysmith, VA 22501, USA). The computer game linked to the manipulandum provides motivational and quantitative feedback. Courtesy of Biometrics Limited, Ladysmith, Virginia.

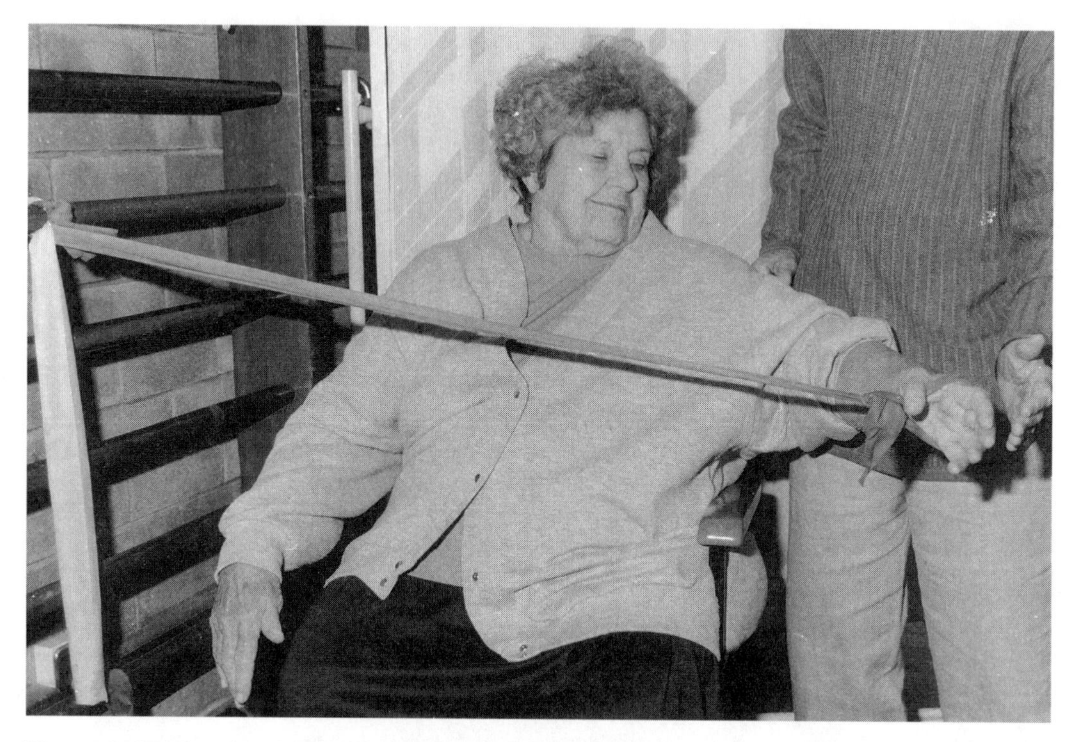

Figure 2–17 Theraband (HCM-Hygenic Corp, Malaysia SDN BHD) of different levels of resistance is used to enable independent practice. In this figure, exercising is directed in particular at glenohumeral external rotator and extensor muscles.

weakness/paralysis, is a major factor in the development of shoulder pain following stroke.[131,146] A program of positioning with prolonged passive stretching of muscles at risk of becoming short because of their resting position (see Figure 2–11) can be instituted early.

- Task-related exercise and training, using a variety of objects to strengthen muscles and improve control and therefore skill (Figures 2–18, 2–19, 2–20). It is likely that active exercise, if sufficiently varied, will itself preserve the muscle's contractility and extensibility. Even if a patient has limited control over the shoulder, exercises and task practice can occur while sitting and standing with the arm supported on a table. Exercise and practice involve bimanual actions as well as unimanual because a person will only have a functional upper limb if the affected limb can be used in conjunction with the other.
- Intensive practice of a variety of tasks with different objects is necessary to train flexibility of performance. Emphasis should be on moving objects around in the hand as well on grasping and holding objects.

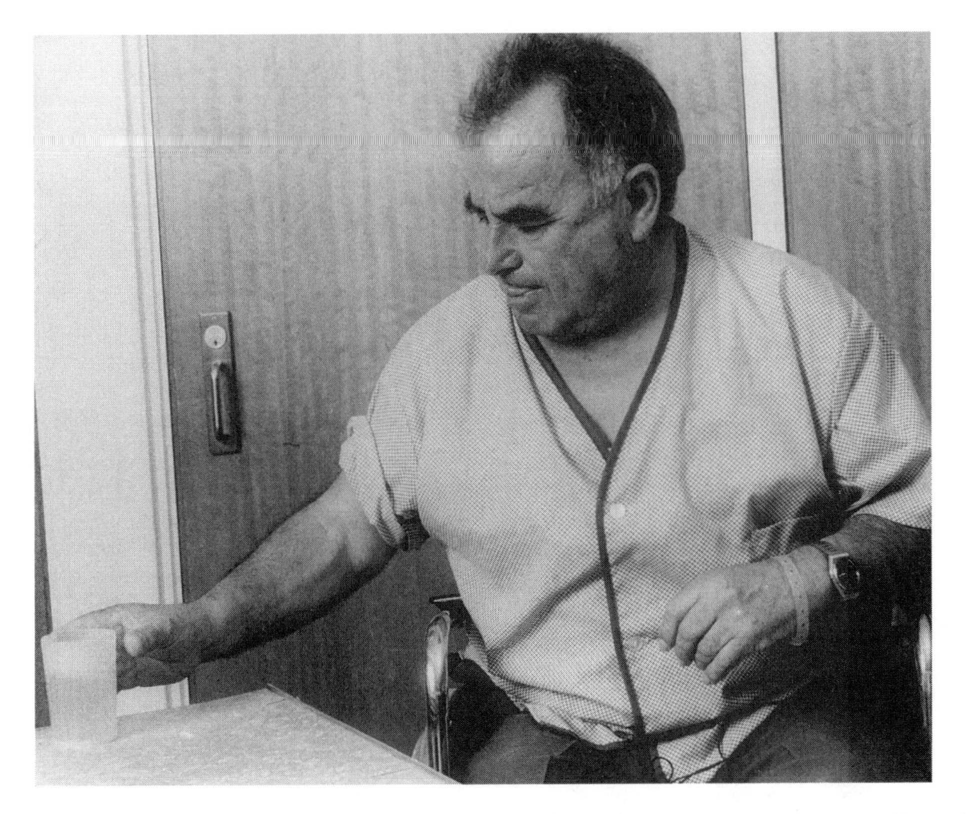

Figure 2–18 Reaching to pick up a glass. Placement of the glass requires more external rotation at the glenohumeral joint than if the object was directly in front.

- All of the points listed above will help to prevent what Taub called "learned non-use."[30] Constraint of the *unaffected* limb and forced use of the affected limb appears to be an effective method of promoting use of the limb.[10,147] It is not yet known whether a patient with total limb paralysis would benefit from this procedure. Taub and colleagues included in their study only patients who could extend their fingers at least 10° and their wrist 20°. Until there is more investigation of this methodology it may be advisable to follow Taub and colleagues' protocol.[10] From the evidence regarding the effects of early vigorous training in rats (see Chapter 4), it may be advisable to avoid constraining the unaffected limb with forced use of the affected limb in the acute stage following stroke. However, the optimal timing of constraint needs investigation in the clinic.
- Restraint of the *hemiparetic* limb (eg, in a sling) should be avoided because this will encourage adaptive muscle shortening and stiffness and learned

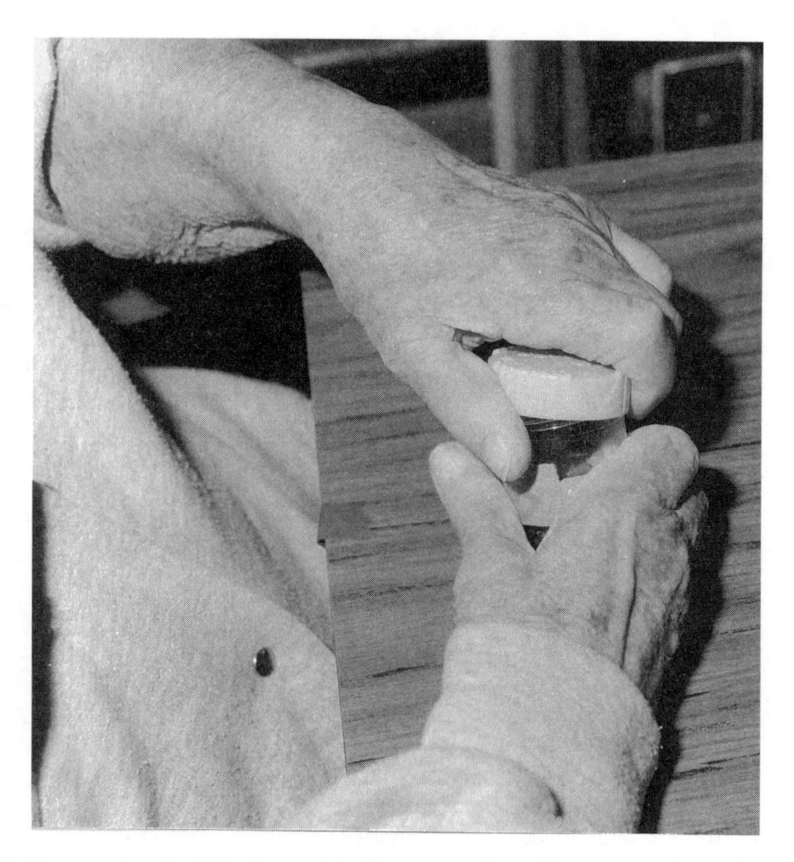

Figure 2–19 Unscrewing the lid of a jar. This exercise provides practice of linking the action of both limbs.

non-use. No evidence shows that the use of slings has any beneficial effect, yet their use appears to be mandatory in many rehabilitation units.

Skill Acquisition

We first suggested in 1980[148] that principles of motor learning arising from research involving healthy persons could be used to improve motor performance in individuals with movement dysfunction. More recent clinical research applying this knowledge appears to support the authors' original hypothesis that motor learning should form a basis for the development and testing of a scientific framework on which to base clinical practice. In Chapter 3 a comprehensive review of motor learning theory and research is included. Some clinical examples of applying this information are provided in the following section.

Stages of Learning

When training patients to optimize motor performance Fitts' three stages of learning[149] can provide guidance. The early or cognitive stage is the acquisition of factual knowledge before attempting any action. The patient needs information about the goal of the action and some idea of how the goal is to be accomplished. In the cognitive stage, the learner is getting the idea of what to do and how to do it, eg, where to put the feet to stand up. Performance during this stage is marked by a large number of errors and, although patients may be aware of doing something wrong, they may not know what they need to do to improve performance. The learner also engages in cognitive activity while listening to instructions and receiving feedback from the therapist as coach. It is important in acute rehabilitation, however, to recall that—as Salmoni[150] points out—for skill learning to occur the person has to move actively.

It is during the intermediate or associative phase that the learner starts to diminish errors that are inherent in the performance of a new skill[151] or, in rehabilitation, in the relearning of "old skills" in the presence of impairments.

Figure 2–20 Removing the lid from a can requires different grasping and manipulating actions from unscrewing a lid.

Fitts and Posner[151] suggested that during the final or autonomous phase of skill learning, speed and efficiency of performance gradually improve and become less subject to cognitive control and less affected by interference. In the clinical setting, therefore, the environment is organized to incorporate both distraction and, once the individual is performing without conscious thought, the opportunity to practice two tasks simultaneously. People at this stage do not consciously think about what they are doing while performing the skill and can start to do another task at the same time, eg, talking while standing up and walking. Magill[21] suggests that the quality of instruction and type of practice are important factors in achieving this final stage. Note, however, that doing two tasks simultaneously can be quite disruptive for patients having problems with working memory.

Identification of Goal

Learning a motor skill involves two critical components in the early stages: 1) identifying what is to be learned; and 2) understanding the ways through which the goal can be accomplished. The two most common ways to communicate the goal of the action are *verbal instruction* and *demonstration,* both live and recorded. Instructions that are appropriately conveyed can obviously be useful in relaying information concerning the goal, but their effect in conveying information about the actual movements to perform is less clear. Therefore, instructions about the manner in which the goal can be achieved are best kept to a minimum. Certainly, it seems that the more the therapist knows about the task, its kinematics, and dynamics, the better he or she can identify for the patient the most important aspects to concentrate on during performance.

Some goals seem to facilitate action more readily than others. As an example, tasks that have goals directed toward controlling one's physical interaction with objects or persons in the immediate environment (concrete tasks) seem to have more meaning and be more motivating than goals directed at movement for its own sake (abstract tasks). Evidence indicates that individuals with movement dysfunction significantly improve their performance compared with controls when an action is presented as a concrete task as opposed to an abstract task, eg, adults with restricted range of motion in the elbow or shoulder as a result of injury[152] and children with cerebral palsy.[153] Abstract and concrete tasks differ in the degree to which the required action is directed toward controlling physical interaction with the environment as opposed to merely moving a limb. Recent evidence indicates that miming a task, ie, performing a task without the relevant object (eg, reaching as though to pick up a cup without the cup being present) results in different patterns of movement compared with the same task performed with the object.[154]

The goal of the action and the movements to be executed can be demonstrated (or modeled) either live or on videotape. When demonstration is the method used for conveying the goal, the patient is expected to remember the observed movement and to create similar movements. The empirical work on the effectiveness of demonstration, however, has been sporadic and the results equivocal.

It is said that modeling of an action either by the therapist or videotape can help the learner develop a conceptual representation[155] or template[156] of the action with instruction kept to a minimum. Observation seems to assist an individual to learn the temporal and spatial features of an action, and learning has been found to be more effective if both model and subject maintain the same spatial orientation.[157]

The therapist needs some knowledge of the patient's existing idea of the action to be performed. This can be achieved by asking the person to describe what has to be done, eg, to stand up. A goal stated in terms of the relationship of body parts may help the patient organize the movement. For example, in standing up, "Move your shoulders forward and up" provides clues about which body segment to move and in what direction, in contrast to the commonly used phrase "Lean forward and stand up." The goal may also be to complete a certain number of repetitions or keep performing the action for a given period of time. Performance needs to be effective; otherwise, repetition in itself will not necessarily improve skill.

The goal to be achieved should be meaningful and worthwhile to the individual, reasonably hard yet attainable. In addition, the goal should be short term, ie, able to be achieved within a training session, although it will be directly related to another, longer-term goal. The immediate goal may be to elicit activity in critical muscles (eg, wrist extensors), while longer-term goals would be to strengthen these muscles through repetitive exercise, to practice manipulating objects using the wrist extensors as part of the synergy required, and to practice reaching for an object (which involves wrist extensors) because the act of grasping is closely related to the reaching phase.

Feedback

Feedback refers to the use of sensory information for the control of action in the process of skill acquisition.[158] It can be positive or negative, qualitative or quantitative, and it may motivate the learner to persist in striving toward a goal as well as provide information to facilitate achievement of a goal. Knowledge of results (KR) is information related to achievement of the goal of the action, and it is known to be one of the most potent variables in learning.[159,160] If a clear and concrete goal has been identified for the patient, information about the achievement of the goal will be immediately available to the individual although some prompting may be needed. A second type of feedback, commonly referred to as knowledge of performance (KP), provides information about how the action was performed. The most effective KP seems to be information about critical components of the action and provides prescriptive information about how to correct the errors.[21]

Both KP and KR can be augmented by the therapist verbally through demonstration and through use of instrumentation (eg, videotape, forceplate system), and knowledge gained can be used to correct errors to improve subsequent performance. As long ago as 1927, Thorndike[161] showed that practice with right/wrong feedback was more effective in bringing about learning than practice alone. His subjects, with vision obstructed, drew 3-, 4-, 5-, and 6-inch (5-, 7-, 13-, 15-cm) lines and the experimenter said "right" if the line was within a tolerance band around the correct length, or "wrong" if it was not. The subjects who had practiced with feedback improved their percentage of right scores from 13% at pretest to 55% in the final session (after 4200 lines were drawn). When subjects had no feedback, the percentage of correct lines did not change after 5400 lines were drawn. The findings of Trowbridge and Cason[162] supported Thorndike's work and added another dimension to it. They showed that quantitative feedback about the extent of error produced faster learning than qualitative feedback such as "right" or "wrong."

Given that accurate feedback is essential to learning, the therapist should confine the use of positive reinforcement, eg, "Good," to situations in which it will encourage the individual to repeat a successful or nearly successful performance and not as a reward for a good try. The physical therapist can use feedback as motivation by using phrases such as "You are doing much better/that was a good try" to encourage the person to persist.

Augmented quantitative feedback or biofeedback uses instrumentation to provide visual or auditory feedback that gives information related to some aspect of performance or a physiological process of which a person is not normally aware. Biofeedback is probably not a substitute for motor training but an adjunct. The rationale for such augmented feedback is that an increase in the amount of information available to the individual leads to an increase in learning and to improved performance.[163] The results of investigations of efficacy are, however, equivocal.

The potential of electronic devices in motor training has not yet been realized, perhaps partly because of the continued emphasis on "handling" techniques rather than on the patient as an active participant in exercise and training. Future developments of electronic devices to provide feedback and motivation (including interactive computer games) should enable patients both to enjoy their exercise program and to spend a greater proportion of the day exercising relatively independently.

Although getting the idea of the movement in the early stages of learning may require more immediate and frequent feedback, the therapist should gradually withdraw augmented feedback enabling the individual to use naturally occurring feedback mechanisms to which his or her attention may need to be drawn. This gives the person the opportunity to increase problem-solving ability.

Withdrawing augmented feedback, whether provided by a therapist or electronic device, is not, however, followed necessarily by a continuation of the desired behavior. A neglected issue in the debate on augmented feedback may be in relation to the timing of muscle activity because it is difficult to deliver the biofeedback signal at the appropriate time in the action.[2]

Making Errors

The experience a person has in correcting errors is especially important in skill learning.[21,164] There is a continuing controversy about the role of augmented feedback in terms of whether such feedback should emphasize errors made or aspects of performance that are correct. Magill[21] writes that augmented feedback as error-information functions in an informational role that is related to facilitating skill improvement. Augmented feedback about what the person did correctly has a more motivational role. However, Magill cautions that focusing on what was done correctly is not sufficient to produce optimal learning, particularly in the early phase.

When an individual with movement dysfunction has great difficulty in performing an action and can only attempt that action by making gross errors, *manual guidance* and/or modification of task or environment (Figure 2–7; Figure 2–21) are used either to enable some part of the action to be carried out or to give the person the idea of the movement.

In the skill acquisition literature, two types of manual guidance used in training are referred to as *passive movement* and *spatio-temporal constraint or physical restriction.*[165] Passive movement may involve placing a limb in a position that enables movement to take place or moving a limb to give the person the idea of the movement (ie, to present a model of the goal and the spatial characteristics of the action). Spatial and temporal constraint involves holding part of a limb stable to constrain the action spatially while the patient has only to control part of the action. This cuts down the degrees of freedom that need to be controlled by the individual, allowing for more concentration on a particular problem of muscle activation associated with achieving the goal. As the person develops some control, the therapist decreases the physical constraint and manual guidance is replaced by verbal and object-mediated guidance. This gives the patient the opportunity to learn to control the temporal as well as spatial aspects of the action.

Newell[166] suggests that manual guidance of the second type in the initial stage of learning provides some benefits to the learner because it reduces the likelihood of the learner making errors and thus developing bad habits.

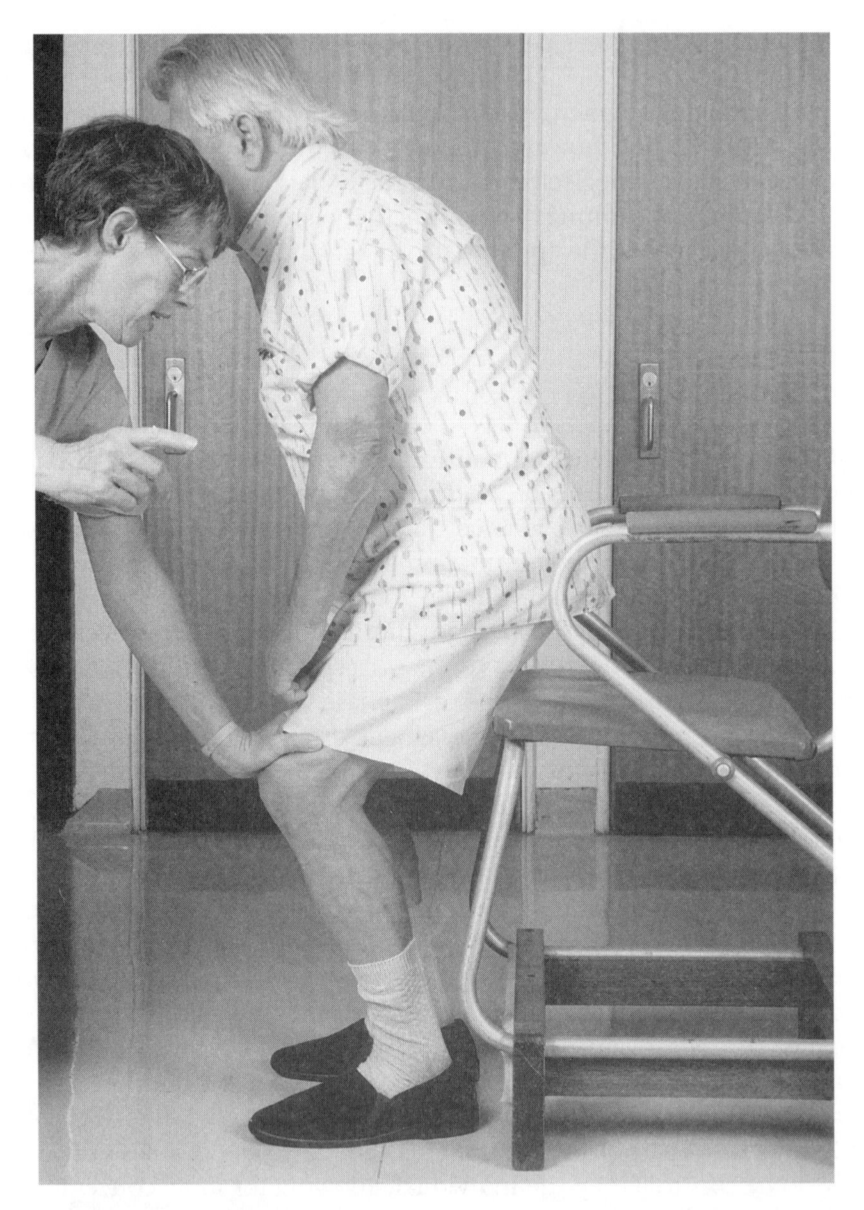

Figure 2–21 Standing up. Stabilizing the foot on the floor, with pressure down along the line of the shank, ensures that knee and hip extensor muscles propel the body mass vertically. The manual pressure ensures weight bearing through the left leg. The seat modification (increased height) reduces the extensor force requirements.

However, he also points out that provided fundamentally inappropriate actions (ineffective adaptive movements) that may hinder learning and create bad habits are prevented, errors of execution (ie, errors in the spatio-temporal details of an action) may actually be beneficial to skill learning.[21] It is only by attempting to carry out a movement to achieve a goal that a person knows whether the action attempted is actually successful. It has also been found in able-bodied persons that motivation is low when they cannot make mistakes during practice.[167] In other words, learning requires an element of trial and error.

It is evident, therefore, that there are different ways of constraining an action to enable an individual/the learner to use available muscle activity as effectively as possible. The therapist may use manual guidance to constrain an action by securing a segment, eg, by aligning the shank segment and securing the foot to the floor to assist the contribution of the hemiparetic leg as the individual practices standing up (Figure 2–21). A splint or strapping can be applied to constrain muscle action and prevent unnecessary joint movement, eg, strapping to the posterior knee prevents excessive extension at the knee while enabling available muscle activation to control lower limb movement in standing and walking (Figure 2–22).

The issue of trial and error in motor learning has considerable implications for patients as learners. It is probable that physiotherapy approaches that advocate limiting practice of actions to avoid errors or "abnormal strategies," need re-evaluating. For example, when training gait, therapists may spend more time on "preparatory" activities[38,168] and be reluctant to encourage patients to walk for fear of "stereotyped mass synergies" (see Hesse et al.[5] for comment).

Therapist-guided movements ("handling"), other than for the purpose outlined previously, should probably be kept to a minimum. Therapists should appreciate the effect placing their hands on a patient has in imposing temporal and spatial details on the patient's performance, and in preventing them from seeing the results of their own attempts at performing the intended task. It is also evident that with a limb held firmly and guided through a task a tendency exists for the therapist's guidance to turn an active attempt at movement by the patient into a passive movement, for the muscle activity to change from eccentric in one muscle group to concentric in the antagonist group, or for the subject to "switch off" and not attend to the task at hand.

Another method of modifying the task and/or environment does not involve manual guidance to prevent errors but rather a simplification of the task or an alteration of environmental features to take into account muscle weakness and lack of control of postural adjustments or limb movement. For example, raising seat height, by decreasing muscle force requirements, may enable a person to practice standing up with weight more evenly distributed between the two lower limbs, ie, without making adaptive movements that favor the hemiparetic leg.

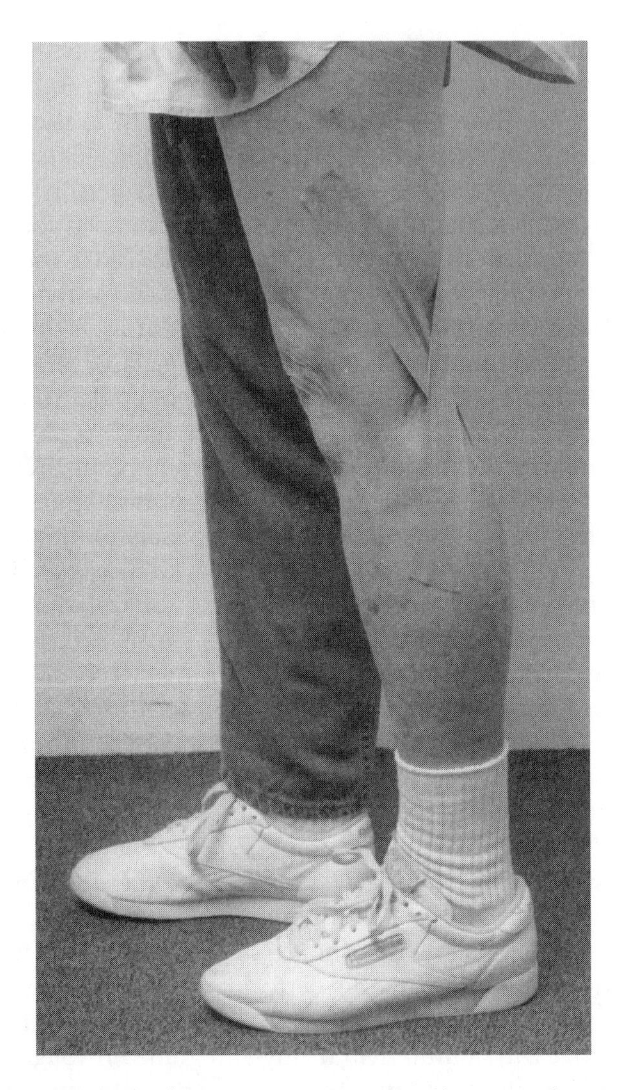

Figure 2–22 Strapping constrains knee movement, preventing excessive extension at the knee.

Practice

As a general rule, skill in performance increases as a direct function of the amount of practice. Johnson[167] defines the law of practice in which the logarithm of the time taken to perform a task plotted against the logarithm of the number of trials produces a straight line. This law holds for the learning of many motor and problem-solving tasks.

Practice can be considered as a continuum of procedures from overt practice at one end of the spectrum to covert or mental practice at the other.[167]

Although the most meaningful and effective training sessions may be those the individual spends with the coach or therapist, both in one-to-one and in group work, it is important in rehabilitation—to ensure the necessary amount of practice takes place—to structure a practice program for the rest of the day.

Mental practice, mental rehearsal or visualization of a motor skill, involves individuals spending time thinking about or imagining themselves performing the movement or action. The relationship between motor and mental activity has been described by several authors.[169,170] Johnson,[169] in a series of experiments that investigated the effect of imagery on performance in able-bodied persons, concluded that real and imagined movements might be functionally equivalent.

Mapping of brain activity during passive observation of someone else's hand and while imagining the grasping of objects with one's own hand has shown different patterns of activity for each mental activity despite no movement having occurred.[171] When subjects were observing hand movements, activation occurred not only in visual cortical areas, but also in areas involved in motor behavior (eg, basal ganglia and cerebellum).

Mental practice may only be suitable for patients who do not have communication or perceptual problems and are able to understand the idea. Time should be taken to ensure that the patient understands the details of the movement to be rehearsed mentally. It is possible that mental practice might have a particular role to play in motor training when the patient has very little muscle activity.

Repetition is an important aspect of practice, and repetition of a task or an exercise has been shown to improve performance in individuals who are disabled as well as in able-bodied individuals.[138,172] It is not clear with individuals with disabilities how many repetitions of an action are necessary to promote improvement. Thousands of repetitions may be necessary to improve performance to an effective level of skill.[173–175] Repetitive practice may be necessary for learning to occur because repetitions enable the system to coordinate the muscular synergies that move the segmental linkage in the desired manner to accomplish the goal.

Repetition can be described as repeated attempts to solve a goal-related problem by building on previous attempts, ie, *repetition without repetition.*[31] Whiting[176] has commented that acquiring skill does not only mean to repeat and consolidate but also to invent and to progress. Thus skill acquisition involves the formulation of new strategies of action and the progressive refinement of solutions to motor problems posed by the environment.

Although investigation into the relative effectiveness of consistent versus variable practice supports the prediction that variable practice—in the able-bodied population at least—leads to better performance than consistent practice, some opposite findings have occurred (for review see references

167,177,178). It appears, therefore, that individuals with movement dysfunction require the opportunity to practice under a variety of relevant situations (walking in a busy corridor, in and out of elevators, on and off a moving walkway). However, when muscle weakness interferes with performance, making movement either impossible or ineffective, repetitive exercise under more stable conditions is probably critical to get the muscles to contract and generate force and, by providing resistance, to increase strength (Figure 2–23).

Although repetitive performance of an action is critical to increasing muscle strength and developing skill, repetitions may be avoided in the clinic because patients with muscle weakness and low endurance fatigue easily. The issue of *fatigability* is, therefore, of concern to therapists organizing exercise and training protocols as well as to physiologists interested in the mechanism by which training induces increases in strength. A study designed to investigate the role of fatigue in strength training compared a training protocol in which able-bodied subjects rested between contractions with one in which they did not rest. The findings showed a substantially greater increase in strength when subjects exercised without rest, at least in the short term.[179] The authors suggested that processes associated with fatigue contributed to the strength-training stimulus.

These results are of interest because—following acute brain lesions—patients may not have the opportunity to carry out sufficient repetitions and have too many rests between trials to enable an increase in muscle strength or to optimize learning. The reason often given is that the patient may experience fatigue. The dilemma is of course that the less people do the more fatigue they are likely to experience. We would point out that provided the person's medical condition is satisfactory, repetition is critical to a successful training program and that patients need to understand that a degree of fatigue (principally muscle fatigue) is normal after exercise. In the clinic as elsewhere fatigue can stem from muscle weakness, loss of physical fitness, and depression. Fatigue is a natural phenomenon as a result of intense (for the individual) physical and mental effort. The fatigue that accompanies physical activity responds to a period of rest, and fatigue within normal limits does not affect learning, although it may temporally affect performance.[180,181]

Following acute brain damage, individuals are known not only to have low endurance to exercise but also have been found to be deconditioned after discharge from rehabilitation.[182–185] The benefits of a physical conditioning program include decreased resistance to fatigue, improved sleep patterns, increased self-esteem, and enhanced participation in other activities. Physiologically, the effects of such programs have been reported to include significant improvement in maximum oxygen consumption, workload, and exercise time.[56,182,183] Potempa and colleagues[56] also found a significant relationship between improvement in aerobic capacity and improved function on the Fugl-Meyer

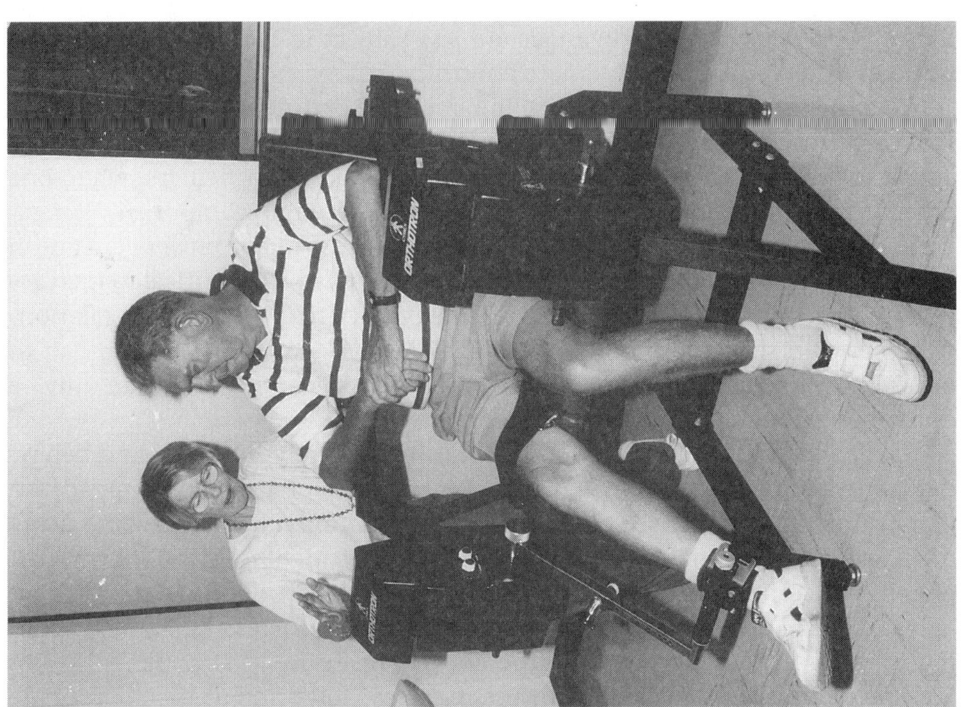

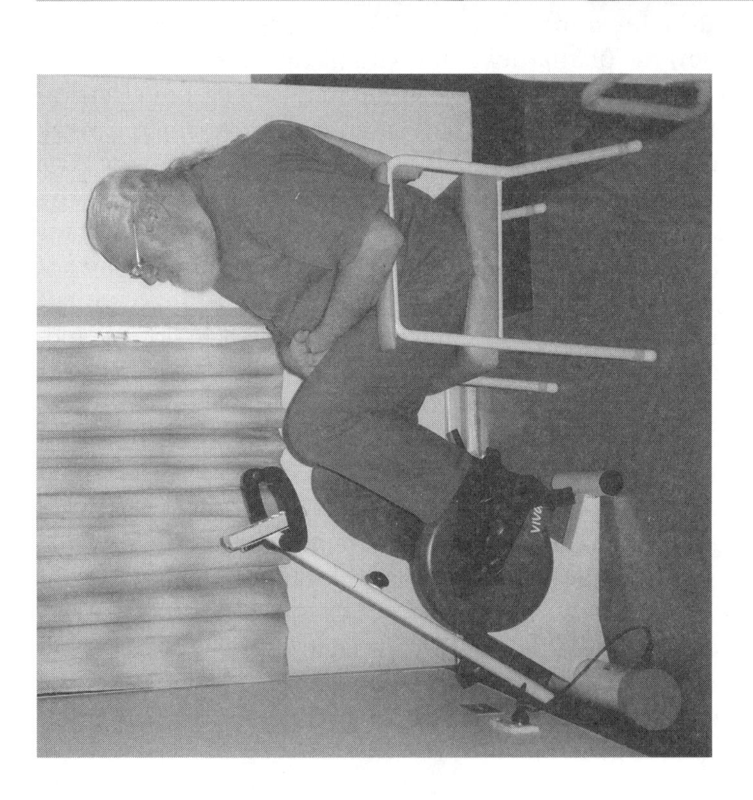

Figure 2–23 Independent exercising with isokinetic machines: (L) The Motomed provides resistance or assistance in response to the patient's performance (Reck, Reckstrasse 1-3, D-88422 Betzenweiller, Germany). (R) Exercising on the Orthotron to strengthen knee extensor muscles concentrically and eccentrically or isometrically (Cybex Human Performance Rehabilitation, 2100 Smithtown Ave, Ronkon Koma, NY 11779, USA).

Index.[186] It appears that individuals are less subject to fatigue when they are stronger, fitter, and, probably, more cheerful.

Another issue in the motor learning literature pertinent to rehabilitation is that of *part versus whole method of practice*. As a general rule, it seems that the action to be acquired should be practiced in its entirety, particularly when one component is to a large extent dependent on a preceding component. For example, several studies of gait have demonstrated that major power generation in the ankle plantarflexors at the end of stance for push-off is critical to increase walking velocity and ensure effective energy patterns.[187,188] Timing relationships between segmental rotations are critical to effective and efficient gait to improve the exchange between kinetic and potential energy[189] and can only be optimized through the practice of walking itself.

There are many reasons, including mechanical, physiological, and behavioral, why functional improvement in performance of an action depends on practice of the action itself. In rehabilitation of individuals following acute brain lesions, however, it may be necessary for the patient to spend time in part practice, attempting to activate paralyzed muscles and/or doing exercises that are specifically directed at strengthening muscles (ie, increasing force generating capacity). Following on from the above gait example, if patients cannot activate plantarflexor muscles with the necessary force and at the appropriate time and range of ankle joint movement, specific exercises are necessary to strengthen (and lengthen) the plantarflexor muscles (Figure 2–24).

Methods of activating a muscle or muscle group may range from electrical stimulation to gain a twitch to practicing part of an action in which a weak muscle or muscle group may be enabled to contract as part of a synergy. The decision regarding part or whole practice becomes a particular issue once the person can generate some force. We agree with Magill[21] that an important theoretical question in part practice is determining which component parts of the skill to practice separately. As discussed previously, we have proposed that critical kinematic features (observable to clinicians) provide a list of component parts to which the patient's attention can be drawn in practice. For example, in the stance and swing phases of walking the essential kinematic components[24] are:

Stance Phase

- Extension at the hip (with dorsiflexion at ankle) to move body forward
- Lateral horizontal shift of pelvis to the stance side (approximately 4–5 cm in total)
- Flexion at the knee (approximately 15°) initiated on heel contact, followed by extension in mid-stance, then flexion before push-off
- Plantarflexion at heel contact, followed by dorsiflexion then plantarflexion at the ankle for push-off

Figure 2–24 Calf muscle exercise to increase strength and to preserve length of the calf muscles (lowering the heels—lengthening contraction of plantarflexors, raising them to plantigrade—shortening contraction). This exercise trains the muscles specifically at the length required at push-off.

Swing Phase

- Flexion at the hip
- Flexion at the knee
- Lateral pelvic list downward (approximately 5°) toward swing side in horizontal plane at toe-off
- Rotation of pelvis forward on swing side (a few degrees)
- Extension at the knee plus dorsiflexion at the ankle immediately before heel contact

Certain major functions, according to Winter,[190] must be performed in effective gait. In the *stance phase* of gait, the lower limb's principal functions are:

- *Support*: the upper body is supported by the prevention of lower limb collapse
- *Balance*: maintenance of upright posture over the base of support
- *Propulsion*: generation of mechanical energy to enable the appropriate forward motion of the body
- *Absorption* of mechanical energy for both shock absorption and to decrease the body's forward velocity.

In the *swing phase*, the lower limb's main function is:

- *Foot clearance*: the lower limb clears the foot for its landing on the support surface, the foot moving on a smooth path through swing from toe-off to heel contact.

In general it can be said that intensive practice is required to:

- Enable a movement pattern to be learned (ie, to refine the neural commands to muscles[191])
- Strengthen muscles
- Train flexible performance, ie, to enable a stable pattern to be modified as necessary according to environmental and other demands.

Unsupervised Practice

The issue of how much time is spent practicing motor skills is a critical one for rehabilitation. For an able-bodied person to gain skill in a particular activity, it is necessary to spend considerable time being coached and practicing independently. We would argue, therefore, that it is also necessary for an individual with a disability. Given the experimental evidence, it is timely for rehabilitation to progress from a reliance on one-to-one therapy, with its time limitations, to a model in which the person practices, not only during one-to-one training sessions with the therapist giving instructions, feedback, and assistance but also in groups (Figure 2–25), each individual working independently at specially designed work stations (similar to circuit training) set up with technological and other aids to promote practice of specific actions and strengthening exercises. One or two therapists, with the help of an aide, can oversee several individuals, with relatives and friends helping out if they are willing. Such organization can increase the time the patient spends practicing without substantially increasing the therapist's load and ensures better use of the short time allowed for rehabilitation in many countries.

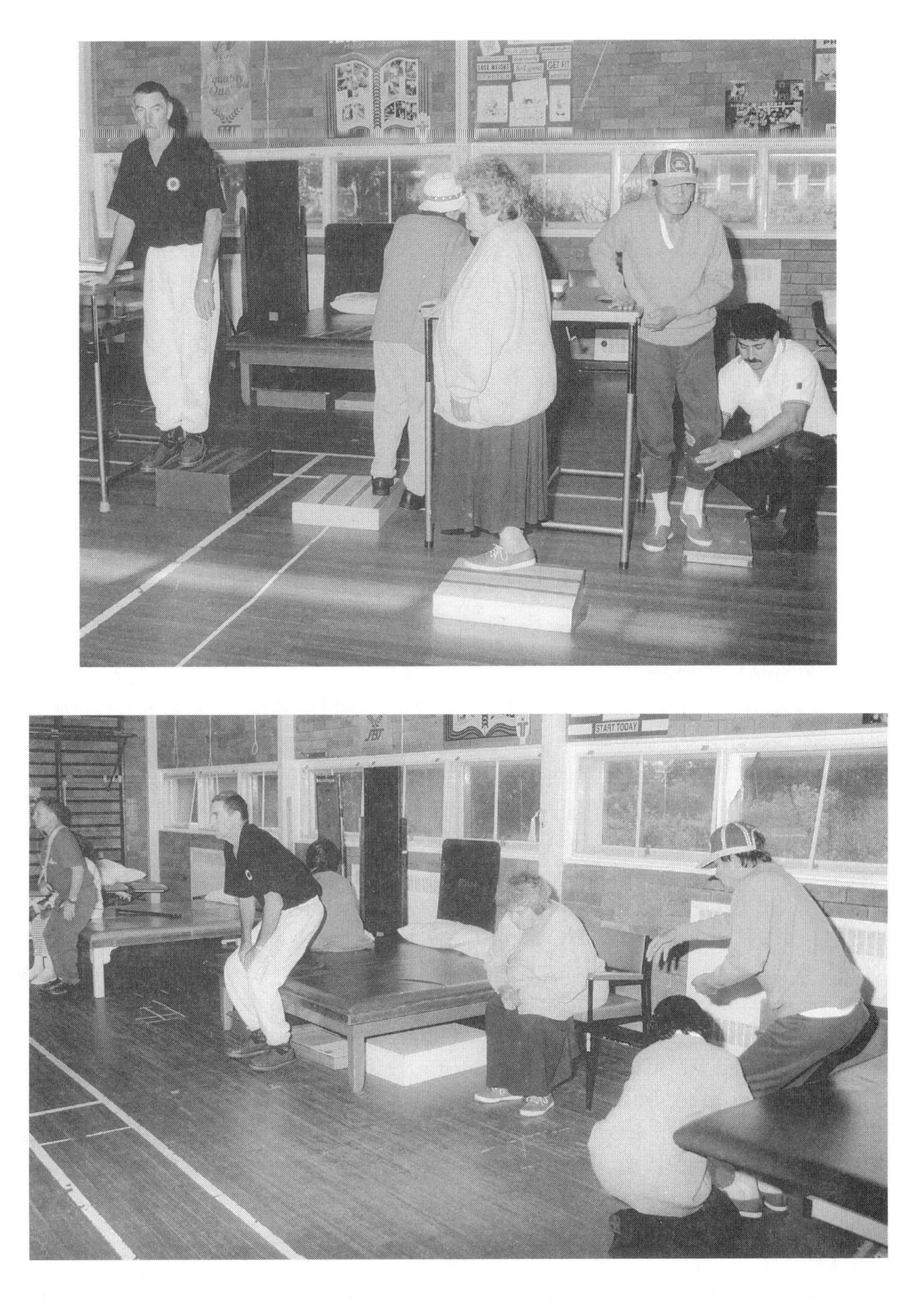

Figure 2–25 Group work. Top: Stepping exercise class. Safety is ensured by having a high table in close proximity for support when necessary. Lower: Standing up and sitting down class. Seat heights are customized according to the individual's lower limb muscle strength. The action is made easier by raising the height of the seat and harder by lowering it.

The work space can be set up to enable task-specific exercise and practice with, for example, treadmill for walking, harness for balancing activities in standing and for walking, electrical stimulation for paretic muscles, practice with electronic feedback and computer-aided devices, and exercise machines such as isokinetic dynamometers.

The therapist provides a performance chart or checklist against which the patient monitors his or her performance. Collection of data to form a learning curve provides feedback and motivation to the patient, relatives, and staff. Periodical anonymous self-efficacy reports seeking the patient's opinions on the benefits of the training program and his or her own sense of well-being and confidence provide information to therapists and assist them to better structure these programs.

Therefore, for intensive, meaningful practice to occur outside individual training sessions the following is required:

- Therapists change their attitudes, particularly their reliance on "one-to-one" and "hands on" therapy.
- The rehabilitation environment is organized to enable safe, independent, relatively unsupervised practice to occur while also providing challenge and motivation.
- Environmental modifications need to be in place to enable the patient to practice particular actions with some degree of success despite impairments.
- Methods of delivery are changed. For example, work stations are set up with written instructions, photographs, videotapes, feedback devices, and exercise equipment to enable circuit training. Written instructions or pictures can provide the specific goals to be practiced, the number of repetitions, and the frequency and distribution of practice sessions. These instructions can be written down by the patient or by the therapist, with the patient verbalizing what to practice.
- Patients are educated to the need to exercise and work hard if the process of rehabilitation is to be optimally successful. In other areas of health research it has been shown that the more an individual understands about his or her condition and the evidence supporting the treatment regimen and the patient's role in it, the better the outcome. This factor does not appear to have been investigated in movement rehabilitation.

Transfer of Learning

Transfer of learning is of obvious importance in skill learning and an issue of particular importance in rehabilitation. In rehabilitation, one of the goals of practice is developing the patient's ability to transfer performance of the skill from the clinical environment to some other environment in which the individual must perform the action (see earlier section on muscle strength specificity).

Some years ago, Thorndike[192] suggested that transfer was the result of identical elements being transferred from one task to another. Since then the motor learning literature on able-bodied persons provides some evidence that a positive transfer can occur to similar tasks.[191,193]

Movements that share similar types of fundamental dynamic behaviors can be considered as particular classes of action. One such class is composed of those actions in which the hips, knees, and ankles, acting as a functional linkage, flex and extend over the feet as a fixed base of support. In these actions, for example, standing up and sitting down, squat-to-stand, stance phase of gait including stair walking, bending down to pick up objects, the lower limb extensor muscles (mono- and bi-articular) contract concentrically and eccentrically to raise and lower the body mass (see Figure 2–2). It is, therefore, possible that simple exercises such as stepping exercises (Figure 2–26) designed to increase the capacity of lower limb extensor muscles to work concentrically and eccentrically to support, raise, and lower the body mass, may lead to an improvement in those actions that are dynamically similar. In support of this concept, practice of stepping exercises by individuals with disabilities appears to transfer into improved performance in gait.[7,16]

Nugent and colleagues[7] reported a relationship between the number of stepping exercises performed and improvement in the Walking Item of the Motor Assessment Scale.[194] Subjects performed a maximum of 60 repetitions of stepping exercises each day with the hemiparetic lower limb raising and lowering the body mass (by extending and flexing the lower limbs) with the foot on a small step (5–12 cm). Using the same exercise, Sherrington and Lord[16] reported a significant relationship between increase in strength and increase in walking speed in a group of elderly subjects, some of whom had had a stroke, after fall-related hip fractures. In contrast, no significant improvements were found in the control group.

A recent study by Buchner and colleagues[68] provides some explanation for these results. The authors investigated the relationship between strength and gait speed in older adults. They found an association between leg strength and gait speed in weaker subjects but no association in stronger subjects. The relationship between leg strength and gait speed was similar for both older men and women. The authors suggested that the findings represent a mechanism for how small changes in physiological capacity may have substantial effects on performance in frail weak adults, whereas large changes in capacity have little or no effect in healthy adults. Buchner and colleagues' investigation supports the view that strength gains carry over into improved functional performance under certain conditions, ie, when subjects already have considerable muscle weakness.

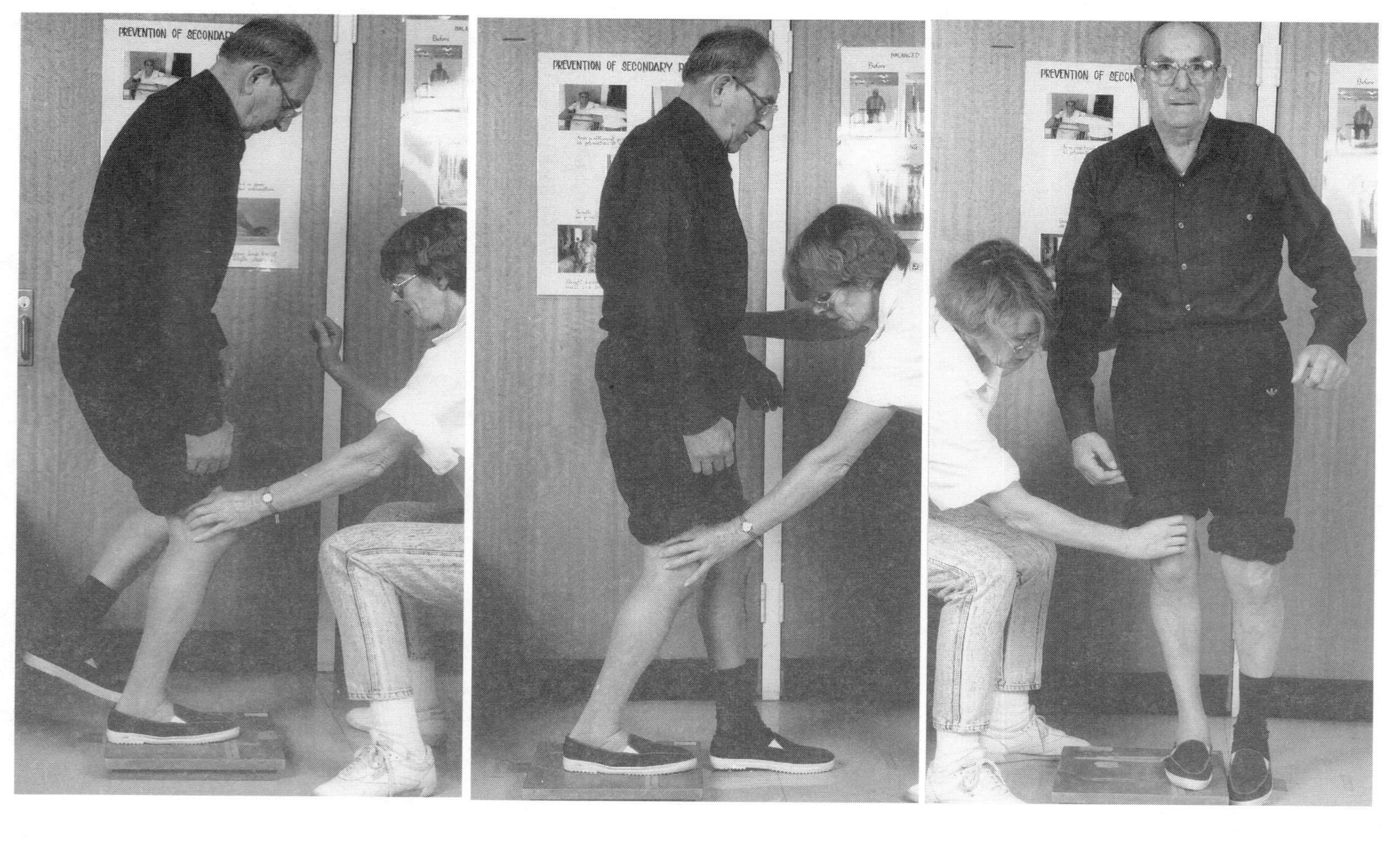

Figure 2–26 Stepping exercises to strengthen lower limb extensor muscles. The exercise provides eccentric and concentric work. From left to right: stepping up; stepping down; lateral stepping. The therapist stabilizes the foot on the ground when necessary. *Source:* Reprinted with permission from L. Ada and C. Canning, *Key Issues in Neurological Physiotherapy*, © 1990.

Apart from a relationship between strength and performance it is interesting to consider what other factors might cause transfer to occur. If these factors are known, then one can understand better what it is about skill that enables the person to adapt his or her performance to a new situation. Magill[21] discusses two hypotheses. Both hypotheses consider the similarity between the two situations to be critical in explaining transfer but differ in the explanation. The first hypothesis proposes that transfer of learning occurs because the components of the skill and/or the context are similar; the second proposes that transfer occurs primarily because of similarities between the amounts and types of learning processes required. Interestingly, the studies reported above by Nugent and colleagues[7] and Sherrington and Lord[16] provide an example of a transfer effect from exercises that comprised similar components to those in the stance phase of gait.

The question still remains as to how to maximize the transfer effect in the clinic. Transfer of learning into different contexts involves in part promoting stability of performance across different environments. This is a difficult point to establish in rehabilitation. To help patients achieve early independence, health professionals may encourage ways of achieving particular objectives that—in the long run—may actually prevent patients from regaining real independence. When a therapist is training walking, for example, it is counterproductive if the person spends the greater part of the day propelling a one-arm drive wheelchair. Although a wheelchair may provide some short-term independence in the sheltered environment of the rehabilitation unit, this activity encourages adaptive movements,[195] exercises the unaffected rather than the affected limbs, probably encourages learned nonuse of the hemiparetic side, and may prevent transfer of learning from occurring.

Similarly, if the therapist is training a patient to stand up, the patient's goal will be to swing the shoulders forward and up while pushing down through the feet, extending knee, hip, and ankle joints to propel the body mass vertically. The patient will, however, only learn to do this and develop skill standing up in a variety of contexts if the opportunity exists to practice intensively and in different environments, ie, outside therapy sessions. Particularly in the early stages the patient will have a natural tendency to stand up using the stronger and more easily controlled intact leg because of difficulty generating force (ie, weakness) in the hemiparetic leg. If nursing or therapy staff reinforce this way of standing up, allowing patients to push down on the arm of the chair and pivot on the intact leg, two methods of standing up are being practiced, and it is likely that the latter method will be learned because it is practiced in a real world context and repeated more frequently.

Practicing sit-to-stand from a higher than average seat reduces the muscle force required,[72] enabling the individual to exercise the leg extensor muscles whenever moving from chair to standing and vice versa. Transfer of learning may, therefore, occur more readily, if "easy to get out of" chairs are also provided in other environments such as in the ward and living areas. The most important factors in chair design to assist rising are the height of the seat, no obstacle to placing the feet back, and the design of the seat. A deep or backward slanting seat increases the distance the body mass has to be moved upward and requires considerable effort to get to the edge of the chair before standing up. Consideration of factors such as these may be critical if transfer of training is to occur.

The Rehabilitation Environment

Training implies that environmental conditions are arranged to produce the most rapid improvement in skill possible. We can consider these conditions from the point of view of the movement organization of the task itself and also in terms of the effect of the rehabilitation environment on the patient as an active participant.

In Chapter 3, issues related to a taxonomy of tasks are reviewed in the context of the environmental conditions in which they take place and the nature of the movement organization. Gentile classifies movements performed under stationary environmental conditions as "closed" and, when the regulatory environmental conditions are themselves in motion, the task is classified as "open." In closed tasks learners are free to perform at their own pace, ie, to select their own temporal organization; they can decide when they start and how long the movement will take (ie, the speed). In contrast, the temporal and spatial characteristics of open tasks are controlled by the temporal and spatial characteristics of the regulatory conditions in the environment, and thus the learner's movements are paced by the movements of other people and objects in the environment. For example, walking down a busy corridor is controlled by its width, placement of doors, and the changing position of people in that corridor as well as the velocity of moving objects and people.

This classification of tasks provides information about the requirements of each task that assist in developing strategies to improve performance. In other words, the therapist needs to analyze not only the task but also the nature of the environmental demands of the task to be learned.

To facilitate optimal practice conditions and increase performance of a particular action in the early stages of recovery, the environment and therefore the task may need to be modified. This could involve, for example, the use of external constraint such as taping to facilitate a muscle or group of muscles to contract and generate force at the appropriate length for a specific action (see Figure 2–22), practicing standing up and sitting down from a higher than average seat to decrease muscle force requirements while still ensuring that the individual is actively participating in performing the task, constraining the intact upper limb to force use of the affected upper limb,[10,196] and practicing walking on a treadmill with supportive harness.[6]

Walking practice in the physical therapy area on a firm, flat surface with the therapist deciding what direction to go and how far is to practice walking in an environment in which conditions are fixed, relatively predictable, and stationary. The patient needs also practice to develop flexibility in matching performance with the spatial and temporal characteristics of the environment, such as walking in a busy corridor, in and out of an elevator, under and over objects; walking in different lighting conditions; and timing movement to external demands, eg, pedestrian traffic lights. The therapist helps the patient develop flexibility in motor performance by providing dynamic practice sessions and by encouraging the patient to search actively for critical environmental cues, particularly visual cues.

With the high cost of providing intensive rehabilitation programs it is critical that optimal conditions are provided for patients to gain functional motor performance as quickly as possible. Currently, the rehabilitation environment may provide protection, safety, and care but lack stimulation, challenge, and the chance to exercise. One outcome of a poor environment may be a mental deterioration that is unrelated to prestroke condition or age but instead results from inactivity, depression, and disuse of mental faculties.[197]

The typical rehabilitation unit has been built to a medical model in which the emphasis is on treatment. The "patient" attends to "receive treatment." As Keith[198] pointed out, referring to a systematic observational study of stroke patients in a rehabilitation hospital, the patient is in an ambiguous position:

> ...on the one hand he is subject to the restrictions inherent in a medically oriented regimen, on the other hand he is expected to participate actively in treatment in order to become as physically and socially competent as possible. A major goal of the organization, to aid individuals toward functional recovery and independence, is often at cross purposes with the mode of service delivery set by professionals in which the patient has limited opportunity to exercise such competence for independence. (p 575)*

In Keith's study, stroke patients were predominantly inactive outside structured therapy time even though observations were taken during the patients' waking day. Although the results of this study were published some years ago, it is disappointing and somewhat surprising that more recent studies spanning several countries confirm the original findings[199–201] and more recently report that patients are actually engaged in motor behavior for as little as 20%–30% of

---

*Source: Reprinted from *Social Science & Medicine*, Vol. 14A, R.A. Keith, Activity Patterns of a Stroke-Rehabilitation Unit, pp. 575–580, Copyright © 1980, with permission from Elsevier Science.

the waking day during weekdays and even less at weekends,[202,203] even when rehabilitation is planned to be intensive. Given that skill in performance is a function of the amount of practice, these results appear to be in conflict with the primary goal of movement rehabilitation that aims to improve motor performance, promote the acquisition of skill, and promote functional independence.

The major theoretical and applied problems investigated in the study of human ecology are how space and objects mediate interpersonal relationships, how environments can be altered to influence behavior, and the effect on the person of such factors as institution size, crowding, and that person's sensitivity or susceptibility to environmental forces. Ittelson and co-workers have suggested that a person's behavior within a specific environment is systematically related to the environment. It follows, therefore, that the environment can be designed to be a facilitator of desired behavior.[204]

Barker described the close-fitting relationship between humans, objects, and actions. He demonstrated that certain behavior is demanded by certain settings.[205] This is a useful point for health professionals because often a tendency exists that attributes behavior more to the person's stroke, personality, lack of motivation, and short attention span than to the effects of the hospital and rehabilitation environment.

If Barker's concept is applied to many rehabilitation units (with parallel bars, mats on the floor, four-point canes, shoulder slings, one-arm drive wheelchairs), it can be seen that the rehabilitation environment may actually mediate against the objectives of rehabilitation. In this special, rather closed setting, patients hope to learn how to cope with a diversity of settings in the outside world. The fact that transfer of learning from therapy room to elsewhere in the rehabilitation unit and to the outside world frequently does not appear to take place may indicate a major flaw in the provision of clinical practice and the environment in which it is conducted.

Willems[206] suggested that, because the main objective is to train patients to be as self-sufficient as possible, therapy programs should be examined to learn why patients are relatively dependent and inactive in them. One reason, he suggests, may be that staff are overly protective of patients. He observed and systematically recorded the behavior of several patients with spinal cord injuries as they went from one hospital setting to another. He found that these patients were more independent and active in some settings, such as cafeteria and hallways, than in others, such as occupational therapy, recreational therapy, and physical therapy.

In a sense the rehabilitation environment does not prepare the patient for life outside the institution. Few demands are made on the patient and with help at hand, food prepared and served, and a wheelchair provided, the person can do those limited things that are required and may therefore lack the motivation and the sense of urgency that would create the need to do more. The patient may have low expectations and may, during the stay in the hospital or rehabilitation setting, become satisfied with a relatively low level of achievement, a lower level than could actually be achieved with more effort and time spent in exercise.

Effects of Use and Experience on the Recovery Process

Parallel to the assumptions about motor control and motor learning that underlie this scientific framework for rehabilitation are assumptions about the recovery of the system from the original cause of the motor impairment. We make the assumption that the person's experiences (including physical therapy) affect recovery either positively or negatively.

All living organisms have an inherent capacity to adapt throughout life, and organizational processes affecting all systems are reflective of the organism's history, including experience and use. With technological advances (transcranial magnetic stimulation (TMS), positron emission tomography (PET), and other functional imaging techniques) the aetiology and mechanisms of organizational processes are increasingly being understood and such knowledge provides insights into how these processes can be manipulated to drive optimal recovery after a CNS lesion. It is reasonable to hypothesize, as we have in several texts,[24,69] that training following a lesion, given that it involves the person learning again how to perform actions and mental processes that were performed with ease prelesion, is a critical stimulus to the making of new or more effective functional connections within remaining brain tissue. Although different connections may mediate the action after the lesion, what seems certain from current research is that for rehabilitation to be effective in aiding an individual to regain optimal functional recovery, more emphasis needs to be placed on methods of "forcing" use of affected limbs through task-specific exercise and training.

Mounting evidence indicates that reorganization reflects patterns of use. Evidence has accumulated over the past decade that motor and sensory cortices are capable of a remarkable degree of reorganization throughout life, after injury to the CNS and after different kinds of manipulations. The evidence of a link between patterns of use and the capacity for functional gain for the individual comes from evidence of reorganization of cortical outputs following amputation in humans,[207–209] of differing cortical maps depending on modification of activity of peripheral sensory pathways in monkeys,[210] and functional changes in cortical motor and sensory neurons in response to training in humans.[211–215] These training studies are of particular interest to physical therapists involved in training patients who need to relearn previously well-learned skills. For example, Pascual-Leone and colleagues,[212] using focal TMS, found that motor representations of the long finger flexor and extensor muscles of the trained hand enlarged after a 5-day course of one-handed, five-finger exercises on the piano. Similarly, right-handed string players have been shown to have a substantially increased cortical representation of the fingers on the left hand in the primary somatosensory cortices.[215] The effect was smallest for the left thumb, which is engaged least of the fingers during string playing, and the extent of the change in the representations correlated with the experience of the individual.

Conversely, decreased use of a body part, such as restriction of volitional movement, also induces alterations of motor representations. For example, a significant decrease in motor representations of inactive muscles was found after 6 weeks of unilateral ankle immobilization of healthy individuals without neurological illness.[216] These changes became even more distinct with longer durations of immobilization. In Chapter 4, Held discusses the evidence for CNS reorganization after brain lesions in response to environmental manipulation, training, and pharmacological intervention.

Although there is increasing acceptance of a link between brain plasticity (ie, anatomical, physiological, and functional reorganization) and recovery, it is still not generally accepted in medical or therapy practice that there might be a link between events occurring post lesion and recovery, in particular, events related to rehabilitation. Many clinicians would argue that rehabilitation is critical to recovery but only a few would take the next step and argue that it is the nature and the process of rehabilitation and the methods used that affect recovery; that some methods in use may actually inhibit recovery. It is possible that the typical picture of a person following stroke, as either wheelchair bound or walking with a foot orthosis and four-point cane, with a stiff useless upper limb, may be—at least in part—the result of events and adaptations subsequent to the stroke and preventable in many patients with more effective rehabilitation methods. In our view, major clinical research emphasis should be placed on studying the effects of different rehabilitation methods on nervous system morphology and function as well as on motor performance in order to establish a process of rehabilitation based on evidence of effectiveness.

Impairments Underlying Functional Disability

Immediately following an acute brain lesion affecting the upper motoneuron (eg, stroke) and for variable periods afterward many patients are unable to activate muscles voluntarily and have muscle weakness on the affected side and varying degrees of insensitivity on that side. Some patients demonstrate other signs of damage to the complex connections between visual, auditory, cognitive, and motor control centers, which are manifested as visuospatial agnosia ("neglect"), dysarthria, dysphasia, visual field deficits, mental confusion, and attentional deficits. In the next few pages certain aspects of the motor control problems most commonly found following an acute brain lesion are briefly described. A fuller description is given elsewhere.[24]

Primary Motor Impairments: Positive and Negative Features

More than a decade ago, in an earlier edition, we were reporting evidence that spasticity may not be the major feature interfering with functional recovery following stroke. Since then increasing evidence provides further confirmation that it is muscle weakness and the adaptive changes to soft tissue (particularly muscle) that are the major obstacles to functional recovery;[217] ie, it is the negative features of UMNL and secondary adaptations that underlie the functional movement disorder. In clinical practice, however, spasticity persists as the major concern.

Figure 2–27 illustrates the current view of the impairments following a lesion of the UMN. The division of impairments into positive and negative features has been in common use in neurology for decades[217–219] and this categorization has some explanatory value for clinical practice.[24] We have proposed a

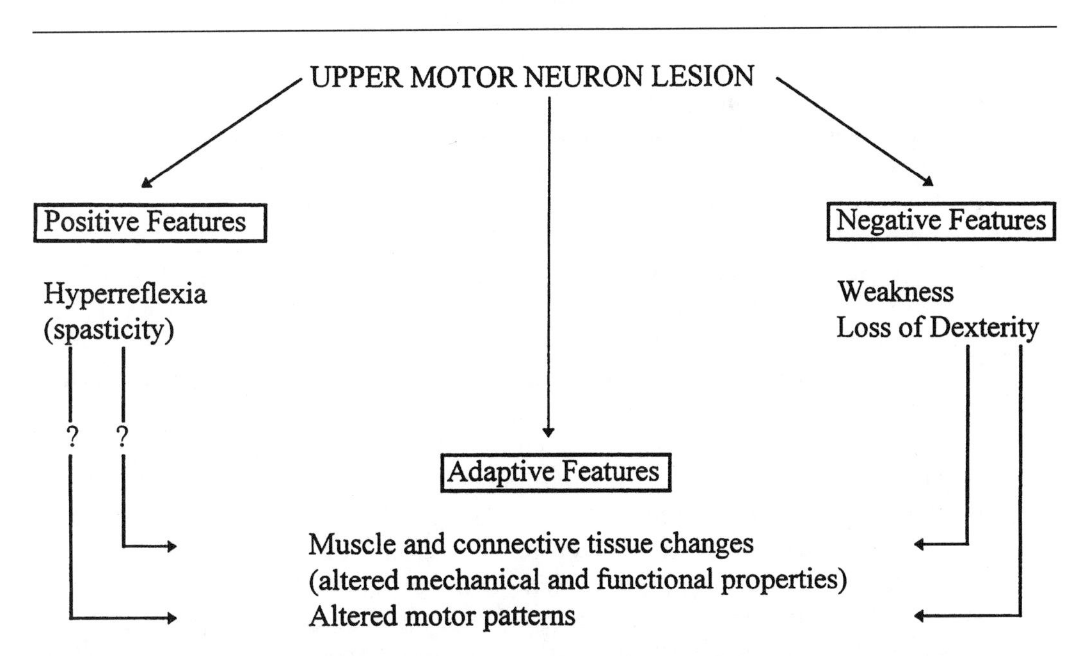

Figure 2–27 The positive, negative, and adaptive features of the upper motor neuron syndrome. *Source:* Reprinted with permission from J.H. Carr and R.B. Shepherd, *Neurological Physiotherapy Optimizing Motor Performance*, p. 158, © 1999, Butterworth Heinemann.

third group of characteristics called adaptive features because it is becoming increasingly evident from scientific studies that adaptive changes to the neural system, muscles and soft tissues, and adaptive motor behaviors form an identifiable set of characteristics occurring secondary to the lesion.

The positive and negative features appear to be relatively independent phenomena related to site and amount of tissue damage and spontaneous recovery processes.[219] Positive features include increased proprioceptive and cutaneous reflexes (hyperreflexia or spasticity) and may be related to secondary functional disturbances in surviving tissue. The negative features, which include weakness, slowness of movement, loss of dexterity, fatigability, and the adaptive changes to muscles are clearly more disabling to most patients than hyperreflexia.

Weakness. The negative features result from insufficient descending inputs converging on the final motoneuron population either to shape complex movements by graded activation of coordinating muscles or to bring motoneurons to the high frequency discharges necessary for tetanic contraction strength.[220] This insufficiency results in weakness (loss of strength of voluntary muscle action), slowness of movement, and loss of dexterity and coordination. Disorganization of motor output at the segmental level also contributes to the weakness.[221]

Muscle force is dependent on the number and type of motor units recruited and the characteristics both of motor unit discharge and of the muscle itself. Muscle force is increased by increasing the number of active motor units and the firing rates of those units. After a UMN lesion, weakness is demonstrated by deficiencies in generating force and sustaining force output.[222] This results from loss of motor unit activation as well as changes in recruitment ordering and in firing rates.[221–223] In addition, adaptive changes occur in the properties of motor units and in the morphological and mechanical properties of the muscle itself as a result of loss of innervation, immobility, and disuse.[223–226]

Slowness of Movement. Slowness of movement and in initiating movement seems to result from difficulty in building up tension during the initiation of movement. It is more evident in fast than in slow movements.[226–228] Several studies of individuals with stroke have reported slower than normal performance of actions such as walking and sit-to-stand.[3,229]

Loss of Dexterity. Loss of dexterity is difficulty making independent movements, particularly for fine manipulation tasks, or lack of adroitness.[217,230] Dexterity appears to depend on the sustained and rapid transfer of sensorimotor information between cerebral cortex and spinal cord.[231]

Reflex Hyperactivity (Spasticity). The positive features are the result of abnormal proprioceptive and cutaneous reflexes,[217] ie, reflex hyperexcitability or spasticity. Considerable confusion exists about the phenomenon involved in what is called spasticity. The term is often used broadly by medical and therapy clinicians to reflect a combination of negative and positive features, hyperreflexia, weakness, abnormal movement patterns, and hypertonus (resistance to passive movement). This lack of clarity persists despite the definition of spasticity agreed to at a neurological meeting and reported by Lance in 1980[232] (and subsequently in 1990[233]) in which spasticity, one component of the upper motoneuron syndrome, was defined as a motor disorder characterized by a velocity-dependent increase in tonic stretch reflexes, with exaggerated tendon jerks resulting from hyperactivity of the stretch reflex.[232]

Spasticity is typically tested in the clinic by passive movement of a limb or joint, by scales such as the Ashworth scale[234] and the pendulum test.[235] Resistance to passive movement or hypertonus is taken as indicative of abnormal excitability of components of the segmental stretch reflex. However, these tests do not distinguish between the peripheral contribution to the resistance, which results from an adaptive increase in muscle stiffness, and the neural contribution resulting from hyperreflexia.[236,237] Muscle "tone" is normally dependent on mechanical rather than neural factors. Any resistance to passive movement in an individual, whether able-bodied or following a brain lesion, is determined by three components: physical inertia of the extremity; mechanical-elastic factors (compliance of muscle, tendon, and connective tissue); and, where present, reflex muscle contraction.[238] Increasing evidence indicates that mechanical and morphological changes in soft tissues play the major role in resistance to passive movement.[230]

Clinicians may also assume that abnormal patterns of movement imply the presence of spasticity.[233] It is very likely, however, that there is an alternate explanation for many patients. The patient with muscle weakness demonstrates adaptive motor patterns ("abnormal patterns") in an attempt to achieve a goal (Figure 2–28). These patterns are imposed on the individual by the effects of muscle imbalance and, frequently, the effects of muscle shortening and stiffness.

It is not mere pedantry that underlies the perceived need to clarify spasticity. A clinician's understanding of mechanisms underlying a clinical sign is directly related to the therapy given. The classification given in Figure 2–27 enables the planning of clinical intervention to be more specifically directed toward the cause of motor deficits. In brief, if muscle weakness has a significant effect on performance, strengthening exercises are given; if muscle stiffness or muscle contracture impedes the necessary range of movement and the muscle force that can be produced, exercises are given that actively stretch the shortened muscles (Figure 2–29) plus passive stretching (Figure 2–30), with a strict protocol for stretching at-risk muscles for periods during the day. In some cases, serial casting is necessary. Similarly, if velocity-dependent hyperreflexia is found to have a significant effect on movement, prior application of baclofen and local techniques such as botulinum toxin injection, may result in more effective exercise and training.

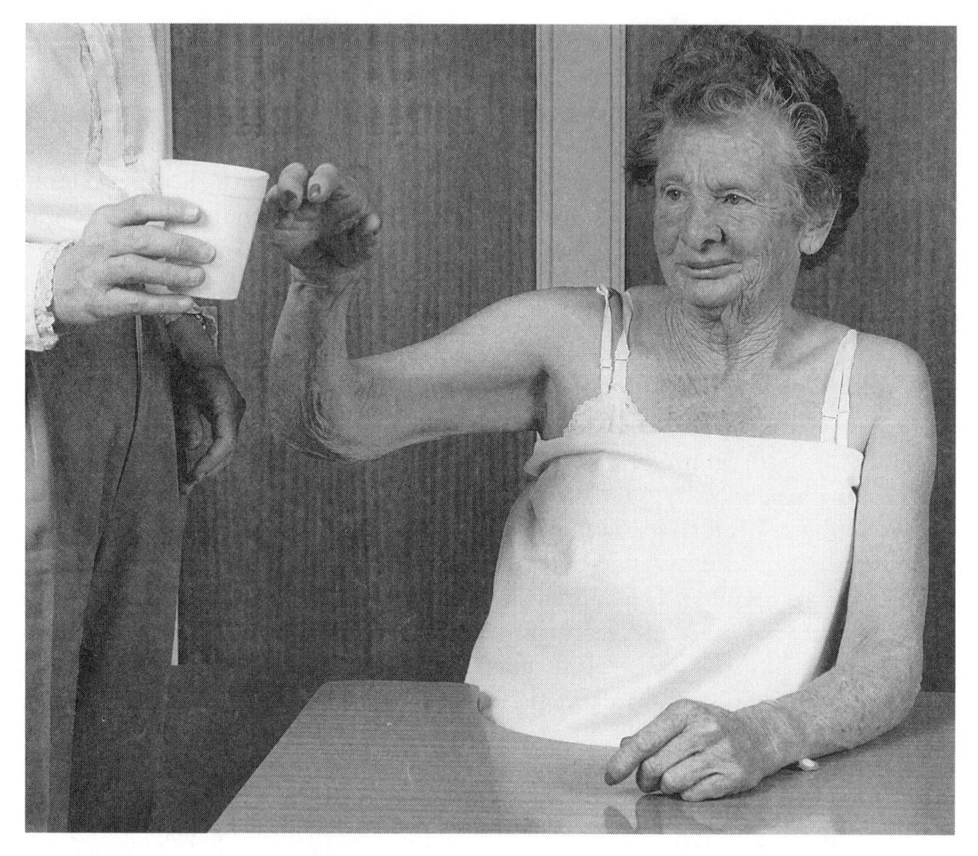

Figure 2–28 The adaptive motor patterns seen here can emerge principally from weakness of shoulder flexor and external rotator muscles and forearm supinators as well as shortening of the internal rotators.

Secondary Impairments: Adaptive Changes to Muscle and Other Soft Tissues

Increasing evidence indicates that *resistance to passive movement (hypertonus)* and the stiffness described by patients result primarily from changes in mechanical fiber properties of muscle and in the tendon,[239–242] which are probably caused by immobility and disuse. Connective tissue stiffness may also contribute to resistance.[243] Muscles subjected to prolonged positioning, ie, rarely exposed to active and passive stretch, are known to undergo changes in crossbridge connections, lose sarcomeres, and become shorter and stiffer.[242,244–246] Similarly, changes to tendons (eg, tendo Achilles) and connective tissue, such as water loss and collagen deposition, may occur.

Muscle inactivity, with persistent maintenance of postures and length changes, can also be associated with changes in sensitivity of spindles, which are activated at shorter lengths than normal.[247,248] Given that the signal for limb position may be provided by the resting discharge of muscle spindles, position

Figure 2–29 The bilateral action of reaching to open cupboard doors provides practice of coordinating the movement of both limbs. This exercise also provides an active stretch to muscles susceptible to shortening.

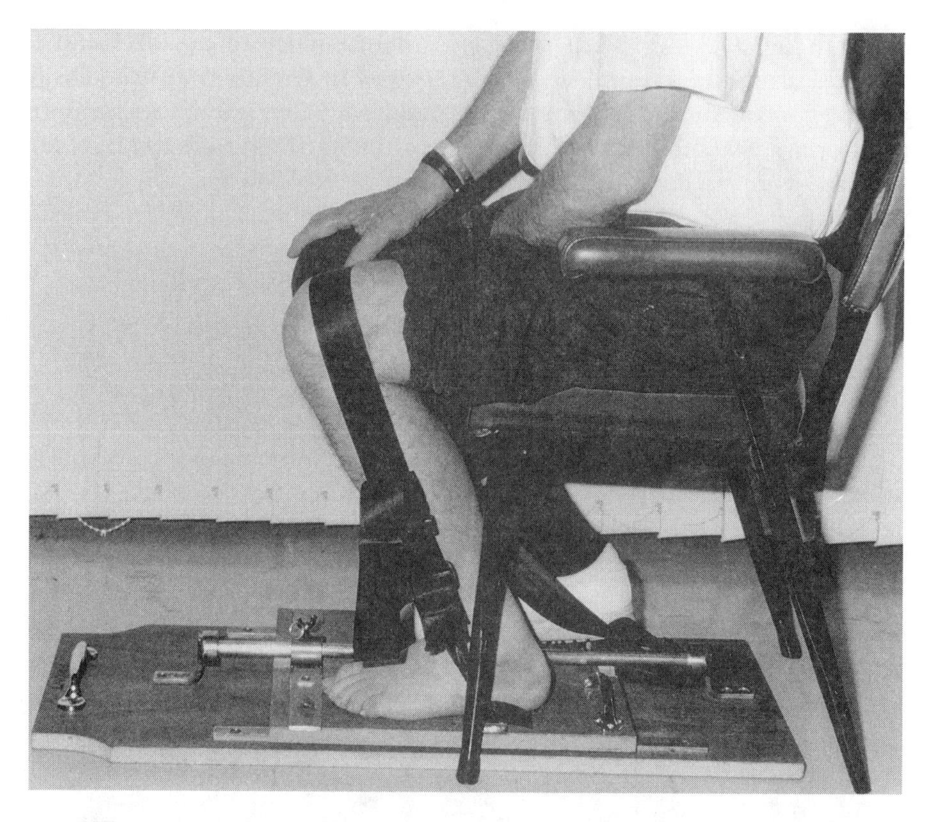

Figure 2–30 This stretcher passively lengthens the soleus muscle. Courtesy of K. Schurr and J. Nugent, Bankstown-Lidcombe Hospital, Sydney, Australia.

sense of a limb may also be affected, which could account for the *altered resting position* of a limb ("abnormal posture") noticeable in many patients following UMN lesion. As Proske and colleagues[249] commented, position sense in a limb may be dependent on the muscle's history, hence, prolonged resting with the elbow in flexion following a stroke would define the "history" of the elbow flexor and extensor muscles. Given that no evidence of a relationship between persistent posturing and hyperreflexia exists, it is likely that persistent posturing is more likely to be related to immobility, disuse changes in soft tissues, and adaptations to resting length of muscles, rather than to spasticity.

Muscle weakness/paralysis, soft tissue changes, and changes in the sensitivity of muscle spindles also underlie the *functional adaptations to motor performance* (adaptive motor patterns) notable following acute brain lesion. In other words, the movement pattern demonstrates the best the person can do under the circumstances, given the state of the neural and musculoskeletal systems and the dynamic possibilities inherent in the multisegmental linkage. The distorted dynamic posturing seen in some patients, however, may result from a different mechanism and is usually called "dystonia."[217] Co-contraction of muscles, often considered indicative of spasticity, can reflect disordered motor control and weakness, and the "stiffening" of a weak and poorly controlled lower limb to prevent it from collapsing.

Somatosensory, Visual, and Perceptual–Cognitive Impairments

Major somatosensory and/or perceptuo-cognitive impairments are found in some individuals following acute brain lesion.[24 for review,250–252] These contribute to poor control of movement and impact on the effectiveness of rehabilitation. Loss of tactile and proprioceptive sensation is considered relatively common, with the most common impairments being in the discrimination and interpretation of information regarding movement (including perception of muscle force), texture, and stereognosis.[253] Carey[39] has pointed out that sensory impairments are often neglected in rehabilitation. However, a body of work is emerging that is relevant to developing strategies for testing and training sensory discrimination and motor performance in clinical practice. The approach to motor training described in this book applies also to sensory training. Challenging and meaningful problems posed to the hand as a sense organ[8] may be solved by the person attending to and concentrating on the sensory inputs and their relationship to the task at hand. The impairment to be overcome is not related to input received from the peripheral system, which is intact, but involves the interpretation of signals distorted by the brain lesion.

Visual impairments, including visual field loss and visual and visuospatial agnosia, also have a negative impact on the ability of people to engage actively with their environment. Perceptuo-cognitive impairments and their behavioral manifestations illustrate the close links between sensory and motor processes and cognitive and perceptual processes. Spatial disorders, for example, can involve impairments of perception, attention, memory, and executive functioning.[254] Lesions of the right brain may be associated with lack of appreciation of spatial aspects of sensory inputs from the left side of the body and left extra personal space. Such individuals do not respond to, identify, or orient toward meaningful stimuli that are contralateral to the lesioned hemisphere.[252] These impairments cause the individual to "neglect" the left side of the body and space (typically called *unilateral spatial neglect or visuospatial agnosia*). Apraxia, a disorder of learned movements not accounted for by weakness, sensory loss, inattention, or lack of comprehension,[255] may also be present.

Measurement

Clinical practice should include the regular collecting of accurate and objective information about the individual's performance of the necessary motor actions. This information drives the design, implementation, and modification of the individual's training program, and, when outcome data from a number of different individuals are collected, such information also provides the evidence that enables the ongoing development of best practice. Since physical therapy is concerned principally with training the individual to improve performance of actions, motor performance is the focus for measurement. Performance measures also provide a test of and information about motor learning.[21]

There has been a reluctance to measure effects of therapy and a continual search for scales that measure the qualitative improvements that therapists perceive. Nevertheless, texts are now available that contain information about the large numbers of valid and reliable tests already available in neurological rehabilitation.[24,256] The tests used depend on the question(s) being asked. Tests available include:

Global measures: eg, Barthel Activities of Daily Living (ADL) Index, Rivermead ADL Index, Functional Independence Measure-FIM
Measures of motor performance: eg, Motor Assessment Scale for Stroke, timed walking, cadence and stride length tests, Functional Reach Test, timed Up-and-Go Test, 9-Hole Peg Test, Spiral Test. Biomechanical measures include kinematics and kinetics, using motion analysis systems, videorecording, forceplate
Measures of muscle function: eg, Medical Research Council grades, hand-held dynamometry, electromyography
Tests of cardiorespiratory fitness: eg, heart rate, 6- to 12-minute walk test
Tests of discrete sensation: eg, Carey's Tactile and Proprioceptive Discrimination Tests
Measures of perception-cognition: eg, Mini-Mental State Examination, Rivermead Perceptual Assessment Battery
Environmental analysis: Behavior mapping, behavior stream analysis
Tests of Memory: Rivermead Behavioural Memory Test

In addition to collecting measurement data, information is collected by observation. This is always open to error and bias but is more reliable if it is structured and if the therapist is well informed with an up-to-date and relevant knowledge base.

Self-assessment and self-efficacy scales and questionnaires provide input from the patient about his or her perception of handicap, quality of life, and ability to perform the necessary tasks at home, work, and leisure. Clinical audits enable a realistic appraisal of the outcome of the rehabilitative process and the contribution of physiotherapy. We agree with Wade[256] that good measures exist for evaluation and that clinicians should collect data routinely and reliably and act on the results of these evaluations.

CONCLUSION

During rehabilitation, the individual with a disability is undergoing reorganization of the neural and muscular systems. We hypothesize that this reorganization, driven by use and experience, can have either a negative outcome or a positive outcome. That the patient is a participant in the movement rehabilitation process is obvious but whether that individual is an active participant depends on the way in which rehabilitation is organized by rehabilitation personnel.

The goal of movement rehabilitation is for the individual to optimize performance and return to many of his or her usual leisure, household, and work-related activities. The regaining of optimal performance appears to depend on a number of key factors that have a high probability of affecting outcome:

- The capacity of individual muscles to generate force and of groups of muscles to work synergistically in the necessary pattern of segmental movement to produce the desired outcome
- The morphology and extensibility of soft tissues (particularly muscle)
- The individual's endurance and physical fitness
- The individual's "drive," understanding of his or her active role, ability to attend to critical features of the environment and the action
- The nature of the rehabilitation environment, whether an active, learning milieu exists in which mental and physical activity occurs, supervised and unsupervised
- The methods used in rehabilitation. There seems to be little realization among health professionals that the methods used may actually make the difference between a positive and a negative outcome. This may underlie the surprising failure to discontinue methodologies shown to be ineffective and that have no rational basis, in favor of methods shown to be more effective.

In regard to the latter two factors, the outcome will be influenced by whether the patient is encouraged to be mentally and physically active; how much practice is expected; what facilities are available to allow stimulating, interesting practice and training sessions; and how effective therapists are in the diagnostic and coaching/training role.

Finally, four basic assumptions underlie our proposals for rehabilitation:

First, that if nondisabled people appear to acquire skill as a result of certain training strategies then it is likely that the field of motor learning can also inform rehabilitation of individuals with disabilities.

Second, that if training is based on biomechanical models of actions, improved performance would be illustrated by a positive change in such biomechanical parameters as angular displacement, velocity, and force production, correlated with positive changes in performance of specific actions. Several studies cited earlier suggest that this is probably true.

Third, that the state of the musculoskeletal system is critical to optimal recovery, so exercise and practice must be active, against appropriate resistance and intensive, to increase strength, preserve muscle length, and increase endurance and cardiovascular fitness.

The fourth assumption is based on the view that the information provided by the context is a major contributor to the optimizing of motor behavior. We are assuming, therefore, that the process of rehabilitation, ie, the goals, methods, and the environmental organization are critical to optimal brain reorganization and optimal motor performance.

We have stressed again in this new edition the need for skill acquisition and physical training in neurorehabilitation. Scientific findings and clinical evidence suggest that the process of rehabilitation should become more active, intensive, and specific to the needs of those individuals with impairments resulting from brain lesions. Understanding the impairments clarifies the needs of particular individuals. However, it is likely that all would benefit from being stronger, better coordinated, more physically fit, and more aware of their own active role in rehabilitation.

REFERENCES

1. Dean CM, Shepherd RB. Task-related training improves performance of seated reaching tasks following stroke: a randomised controlled trial. *Stroke.* 1997;28:722–728.
2. Fowler V, Carr J. Auditory feedback: effects on vertical force production during standing up following stroke. *Int J Rehabil Res.* 1996;19:265–269.
3. Ada L, Westwood P. A kinematic analysis of recovery of the ability to stand up following stroke. *Aus J Physiother.* 1992;38:135–142.
4. Engardt M, Knutsson E, Jonsson M, et al. Dynamic muscle strength training in stroke patients: effects on knee extension torque, electromyographic activity, and motor function. *Arch Phys Med Rehabil.* 1995;113:1459–1476.
5. Hesse S, Bertelt C, Janke MT, et al. Treadmill training with partial body weight support compared with physiotherapy in nonambulatory hemiparetic patients. *Stroke.* 1995;26:976–981.
6. Hesse SA, Jahnke MT, Bertelt CM, et al. Gait outcome in ambulatory hemiparetic patients after a 4-week comprehensive rehabilitation program and prognostic factors. *Stroke.* 1994;25:1999–2004.
7. Nugent JA, Schurr KA, Adams RD. A dose-response relationship between amount of weight bearing exercise and walking outcome following cerebrovascular accident. *Arch Phys Med Rehabil.* 1994;75:399–402.
8. Yekutiel M, Guttman E. A controlled trial of the retraining of the sensory function of the hand in stroke patients. *J Neurol Neurosurg Psychiatry.* 1993;56:241–244.
9. Carey LM, Matyas TA, Oke LE. Sensory loss in stroke patients: effective training of tactile and proprioceptive discrimination. *Arch Phys Med Rehabil.* 1993;74:602–611.
10. Taub E, Miller NE, Novak TA, et al. A technique for improving chronic motor deficit after stroke. *Arch Phys Med Rehabil.* 1993;74:347–354.

11. Wolf SL, LeCraw DE, Baron LA, et al. Forced use of hemiplegic upper extremities to reverse the effect of learned nonuse among chronic stroke and head-injured patients. *Exp Brain Res*. 1989;104:125–132.
12. Hill KM, Harburn KL, Kramer JF. Comparison of balance responses to an external perturbation test, with and without an overhead harness safety system. *Gait and Posture*. 1991;1:27–31.
13. Sunderland A, Tinson DJ, Bradley EL, et al. Enhanced physical therapy improves recovery of arm function after stroke: a randomised controlled trial. *J Neurol Neurosurg Psychiatry*. 1992;55:530–535.
14. Macko RF, DeSouza CA, Tretter BS, et al. Treadmill aerobic exercise training reduces the energy expenditure and cardiovascular demands of hemiparetic gait in chronic stroke patients. *Stroke*. 1997;28:326–330.
15. Moseley AM. The effect of casting combined with stretching on passive dorsiflexion in adults with traumatic head injury. *Phys Ther*. 1997;77:240–247.
16. Sherrington C, Lord SR. Home exercise to improve strength and walking velocity after hip fracture: a randomised controlled trial. *Arch Phys Med Rehabil*. 1997;78:208–212.
17. Blundell S, Shepherd RB, Dean CM, et al. Outcomes of functional strength training in cerebral palsy. In: *Proceedings of 13th International Congress of the World Confederation for Physical Therapy*, Yokohama, Japan: WCPT; 1999.
18. Damiano DL, Abel MF. Functional outcome of strength training in spastic cerebral palsy. *Arch Phys Med Rehabil*. 1998;79:119–125.
19. Jeannerod M. Intersegmental coordination during reaching at natural, visual objects. In: Long J, Baddeley A, eds. *Attention and Performance*. Hillsdale, NJ: Lawrence Erlbaum Associates; 1981:153–168.
20. Iberall T, Bingham G, Arbib MA. Opposition space as a structuring concept for the analysis of skilled hand movements. *Exp Brain Res*. 1986;15:158–173.
21. Magill RA. *Motor Learning Concepts and Applications*. New York: McGraw-Hill International Editions; 1998.
22. Annett J. Acquisition of skill. *Br Med Bull*. 1971;27:266–271.
23. Wilson BA. Management and remediation in brain-injured adults. In: Baddeley AD, Wilson BA, Watts FN, eds. *Handbook of Memory Disorders*. Chichester, England: John Wiley; 1995:451–479.
24. Carr JH, Shepherd RB. *Neurological Rehabilitation: Optimizing Motor Performance*. Oxford, England: Butterworth Heinemann; 1998.
25. Reed ES. An outline of a theory of action systems. *J Motor Behav*. 1982;14:98–134.
26. Gibson JJ. *The Senses Considered as Perceptual Systems*. Boston: Houghton Mifflin; 1966.
27. Lee DN, Lishman JR. Visual proprioceptive control of stance. *Hum Movt Studies*. 1975;1:87–95.
28. Neisser V. *Cognition and Reality*. San Francisco: WH Freeman; 1976.
29. Kelso JAS, Stelmach G. Central and peripheral mechanisms in motor control. In: Stelmach G, ed. *Motor Control: Issues and Trends*. New York: Academic Press; 1976:1–35.
30. Taub E. Somatosensory deafferentation in research with monkeys: implications for rehabilitation medicine. In: Ince LP, ed. *Behavioral and Psychological Rehabilitation Medicine*. Baltimore: Williams & Wilkins; 1980:371–401.
31. Bernstein N. *The Co-ordination and Regulation of Movement*. Oxford, England: Pergamon Press; 1967.
32. Forssberg H, Grillner S, Rossignol S. Phasic gain control of reflexes during spinal locomotion. *Brain Res*. 1977;132:121–139.
33. Fukson O, Berkenblit MB, Feldman AG. The spinal frog takes account of the scheme of its body during the wiping reflex. *Science*. 1980;209:1261–1263.
34. Clark FJ, Burgess PR. Slowly adapting receptors in cat knee joints: can they signal joint angle? *J Neurophysiol*. 1975;38:1448–1463.
35. Kelso JAS, Holt KG, Flatt AE. The role of proprioception in the perception and control of human movement. *Percept Psychophysiol*. 1980;28:45–52.
36. Proske U, Schaible H-G, Schmidt RF. Joint receptors and kinaesthesia. *Exp Brain Res*. 1988;72:219–224.

37. Newell KM. Skill learning. In: Holding DH, ed. *Human Skills.* New York: John Wiley & Sons; 1981:203–226.
38. Ryerson S, Levit K. *Functional Movement Reeducation.* New York: Churchill Livingstone; 1997.
39. Carey LM. Somatosensory loss after stroke. *Crit Rev Phys Rehabil Med.* 1995;7:51–91.
40. Grimby G, Gustafsson E, Peterson L, et al. Quadriceps function and training after knee ligament surgery. *Med Sci Sports & Exerc.* 1980;12:70–75.
41. Rutherford OM, Jones DA, Round JM. Long-lasting unilateral muscle wasting and weakness following injury and immobilization. *Scand J Rehabil Med.* 1990;22:33–37.
42. Sunderland A, Tinson D, Bradley L, et al. Arm function after stroke: an evaluation of grip strength as a measure of recovery and a prognostic indicator. *J Neurol Neurosurg Psychiatry.* 1989;52:1267–1272.
43. Wilson DJ, Baker LL, Craddock J. Functional test for the hemiparetic upper extremity. *Am J Occup Ther.* 1984;38:159–164.
44. Sjostrom M, Fugl-Meyer AR, Nordin G, et al. Post-stroke hemiplegia: crural muscle strength and structure. *Scand J Rehab Med.* 1980;19(Suppl 7),53–61.
45. Bohannon RW. Strength of lower limb muscle related to gait velocity and cadence in stroke patients. *Physio Canada.* 1986;38:204–206.
46. Bohannon RW. Relationship among paretic knee extension strength, maximum weightbearing, and gait speed in patients with stroke. *J Stroke Cerebrovascular Dis.* 1991;1:65–69.
47. Wade DT, Hewer RL. Functional abilities after stroke: measurement, natural history and prognosis. *J Neurol Neurosurg Psychiatry.* 1987;50:177–182.
48. Bohannon RW, Walsh S. Association of paretic lower extremity strength and balance with stair climbing ability in patients with stroke. *J Stroke Cerebrovascular Dis.* 1991;1:129–133.
49. Tinetti ME. Factors associated with serious injury during falls by ambulatory nursing home residents. *J Am Geriatr Soc.* 1987;35:644–648.
50. Lord SR, Clark RD, Webster IW. Postural stability and associated physiological factors in a population of aged persons. *J Gerontol.* 1991;46:M69–76.
51. Frontera WR, Meredith CN, O'Reilly KP, et al. Strength conditioning in older men: skeletal muscle hypertrophy and improved function. *J Appl Physiol.* 1988;64:1038–1044.
52. Fiatarone MA, Marks EC, Ryan ND, et al. High-intensity strength training in nonagenarians. *JAMA.* 1990;263:3029–3034.
53. Posner JD, Gorman KM, Gitlin LN, et al. Effects of exercise training in the elderly on the occurrence and time to onset of cardiovascular diagnoses. *J Am Geriatr Soc.* 1990;38:205–210.
54. Sullivan SJ, Richer E, Laurents F. Programme development: the role of and possibilities for physical conditioning programmes in the rehabilitation of traumatically brain-injured persons. *Brain Injury.* 1990;4:407–414.
55. Lord SR, Castell S. Physical activity program for older persons: effect on balance, strength, neuromuscular control, and reaction time. *Arch Phys Med Rehabil.* 1994;75:648–652.
56. Potempa K, Braum LT, Tinknell T, et al. Benefits of aerobic exercise after stroke. *Sports Med.* 1996; 21:337–346.
57. Milner-Brown HS, Stein RB, Lee RG. Synchronization of human motor units: possible roles of exercise and supra-spinal reflexes. *Electroenceph Clin Neurophysiol.* 1975;38:245.
58. Nativ A. Kinesiological issues in motor retraining following brain trauma. *Crit Rev Phys Rehabil Med.* 1993;5:227–246.
59. Rasch PT, Morehouse LE. Effect of static and dynamic exercise on muscular strength and hypertrophy. *Appl Physiol.* 1957;11:29–34.
60. Sale D, MacDougall D. Specificity in strength training: a review for the coach and athlete. *Can J Appl Sports Sci.* 1981;6:87–92.
61. Rutherford OM, Jones D. The role of learning and coordination in strength training. *Eur J Appl Physiol Occup Physiol.* 1986;55:100–105.

62. Rutherford OM, Grieg CA, Sargeant AJ, et al. Strength training and power output-transference effects in the human quadriceps muscle. *J Sports Sci.* 1986;4:101–107.
63. Westing SH, Seger JY, Thorstensson A, et al. Effects of electrical stimulation on eccentric and concentric torque-velocity relationships during knee extension in man. *Acta Physiol Scand.* 1990;140:17–22.
64. Narici MV, Roi GS, Landoni L, et al. Changes in force, cross-sectional area and neural activation during strength training and detraining of the human quadriceps. *Eur J Appl Physiol Occup Physiol.* 1989; 59:310–319.
65. Kitai TA, Sale DG. Specificity of joint angle in isometric training. *Eur J Appl Physiol.* 1989;58:744–748.
66. Stanish WD, Rubinovich RM, Aurwin S. Eccentric exercise in chronic tendinitis. *Clin Orthop.* 1986; 208:65–68.
67. Tax AAM, Denier van der Gon JJ, Erkelens CJ. Differences in coordination of elbow flexors in force tasks and in movement. *Exp Brain Res.* 1990;81:567–572.
68. Buchner DM, Larson EB, Wagner EH, et al. Evidence for a non-linear relationship between leg strength and gait speed. *Age & Ageing.* 1996;25:386–391.
69. Carr JH, Shepherd RB. *A Motor Relearning Program for Stroke.* 2nd ed. Rockville, MD: Aspen Publishers; 1987.
70. Saunders JB, Inman VT, Eberhart HD. The major determinants in normal and pathological gait. *J Bone Joint Surg.* 1953;35A:543–558.
71. Shepherd RB, Koh HP. Some biomechanical consequences of varying foot placement in sit-to-stand in young women. *Scand J Rehabil Med.* 1996;28:79–88.
72. Rodosky MV, Andriacchi TP, Andersson GBJ. The influence of chair height on lower limb mechanics during rising. *J Orthop Res.* 1989;7:266–271.
73. Pai Y, Rogers MW. Segmental contribution to total body momentum in sit-to-stand. *Med Sci Sports Exerc.* 1991;23:225–230.
74. Shepherd RB, Gentile AM. Sit-to-stand: functional relationships between upper body and lower limb segments. *Human Movt Sci.* 1994;13:817–840.
75. Carr JH, Gentile AM. The effect of arm movement on the biomechanics of standing up. *Human Movt Sci.* 1994;13:175–193.
76. Kreighbaum E, Barthels KM. *Biomechanics: A Qualitative Approach for Studying Human Movement.* 2nd ed. New York: Macmillan; 1985.
77. Cordo PJ, Nashner LM. Properties of postural adjustments associated with rapid arm movements. *Neurophysiol.* 1982;47:287–302.
78. Oddsson L. Coordination of a simple voluntary multijoint movement with postural demands: trunk extension in standing man. *Acta Physiol Scand.* 1988;134:109–118.
79. Zattara M, Bouisset S. Chronometric analysis of the posturo-kinetic programming of voluntary movement. *J Mot Behav.* 1986;18:215–225.
80. Lee WA, Buchanan TS, Rogers MW. Effects of arm acceleration and behavioral conditions on the organization of postural adjustments during arm flexion. *Exp Brain Res.* 1987;66:257–270.
81. Goldie PA, Matyas TA, Evans OM, et al. Maximum voluntary weight-bearing by the affected and unaffected legs in standing following stroke. *Clin Biomech.* 1996;11:333–342.
82. Diener H-C, Dichgans J, Guschlbauer B, et al. The coordination of posture and voluntary movement in patients with cerebellar dysfunction. *Movt Disorders.* 1992;7:14–22.
83. Horak FB, Esselman P, Anderson ME, et al. The effects of movement velocity, mass displaced and task certainty on associated postural adjustments made by normal and hemiplegic individuals. *J Neurol Neurosurg Psychiatry.* 1984;48:1020–1028.
84. Marsden CD, Merton PA. Human postural responses. *Brain.* 1981;104:513–534.
85. Nardone A, Schieppati M. Postural adjustments associated with voluntary contraction of leg muscles in standing man. *Exp Brain Res.* 1988;69:469–480.

86. Do MC, Bussel B, Breniere Y. Influence of plantar cutaneous afferents on early compensatory reactions to forward fall. *Exp Brain Res.* 1990;79:319–324.
87. Winter DA, Patla AE, Frank JS. Assessment of balance control in humans. *Med Prog Technol.* 1990; 16:31–51.
88. Chari VR, Kirby RL. Lower limb influence on sitting balance while reaching forward. *Arch Phys Med Rehabil.* 1986;67:730–733.
89. Dean C, Shepherd RB, Adams R. Sitting balance II: reach direction and thigh support affect the contribution of the lower limbs. *Gait and Posture.* 1999;10:147–153.
90. Nashner LM, McCollum G. The organization of human postural movements: a formal basis and experimental synthesis. *Beh Brain Sci.* 1985;8:135–172.
91. Carello C, Grosofsky A, Reichel FD. Visually perceiving what is reachable. *Ecolog Psych.* 1989;1:27–54.
92. Nashner LM, Cordo PJ. Relation of automatic postural responses and reaction-time voluntary movements of human leg muscles. *Exp Brain Res.* 1981;43:395–405.
93. Nashner LM. Adaptation of human movement to altered environments. *Trends Neurosci.* 1982;5:358–361.
94. Dietz V, Noth J. Preinnervation and stretch responses of triceps brachii in man falling with and without visual control. *Brain Res.* 1978;142:576–579.
95. Lee DN, Aronson E. Visual proprioceptive control of standing in human infants. *Percept Psychophysiol.* 1974;15:529–532.
96. De Wit G. Optic versus vestibular and proprioceptive impulses measured by posturometry. *Agressologie.* 1972;13:75–79.
97. Bobath B. *Adult Hemiplegia: Evaluation and Treatment.* 3rd ed. London: Heinemann; 1990.
98. Davies PM. *Right in the Middle.* Berlin, Germany: Springer-Verlag; 1990.
99. Dickstein R, Heffes Y, Laufer Y, et al. Activation of selected trunk muscles during symmetric functional activities in poststroke hemiparetic and hemiplegic patients. *J Neurol Neurosurg Psychiatry.* 1999; 66:218–221.
100. Colebatch JG, Gandevia SC. The distribution of muscular weakness in upper motor neuron lesions affecting the arm. *Brain.* 1989;112:749–763.
101. Higgins JR. Movements to match environmental demands. *Res Q.* 1972;43:312–336.
102. Shumway-Cook A, Woollacott M. *Motor Control Theory and Practical Application.* Baltimore: Williams & Wilkins; 1995;221–232.
103. James MK, Carr JH, Fowler V. Strength training in chronic stroke patients. In: *Proceedings 13th International Congress of the World Confederation for Physical Therapy.* Yokohama, Japan: World Confederation for Physical Therapy; 1999.
104. Jeannerod M. The timing of natural prehension movements. *J Motor Behav.* 1984;16:235–254.
105. Marteniuk RB, Leavitt JL, MacKenzie CL, et al. Functional relationships between grasp and transport components in a prehension task. *Human Movt Sci.* 1990;9:149–176.
106. Hoff B, Arbib MA. Models of trajectory formation and temporal interaction of reach and grasp. *J Motor Behav.* 1993;25:175–192.
107. Dean C, Shepherd RB, Adams R. Sitting balance 1: Intersegmental coordination and the contribution of the lower limbs during self-paced reaching. *Gait and Posture.* 1999;10:135–146.
108. Kaminsky TR, Bock C, Gentile AM. The coordination between trunk and arm motion during pointing movements. *Exp Brain Res.* 1995;106:457–466.
109. Rosenbaum DA, Vaughan J, Barnes HJ, et al. Time course of movement planning: selection of handgrips for object manipulation. *J Exp Psych: Learning, Memory and Cognition.* 1992;18:1058–1073.
110. van Vliet P. An investigation of the task specificity of reaching: implications for re-training. *Physiother Theory Practice.* 1993;9:69–76.
111. Marteniuk RB, MacKenzie CL, Jeannerod M, et al. Constraints on human arm movement trajectories. *Can J Psychol.* 1987;41:365–378.

112. Weir PL, MacKenzie CL, Marteniuk RG, et al. Is object texture a constraint on human prehension? Kinematic evidence. *J Motor Behav.* 1991;23:205–210.
113. Weir PL, MacKenzie CL, Marteniuk RG, et al. The effects of object weight on the kinematics of prehension. *J Motor Behav.* 1991;23:192–201.
114. Fikes TG, Klatzky RL, Lederman SJ. Effects of object texture on precontact movement time in human prehension. *J Motor Behav.* 1994;26:325–332.
115. Crosbie J, Shepherd RB, Squire T. Postural and voluntary movement during reaching in sitting: the role of the lower limbs. *J Hum Movt Studies.* 1995;28:103–126.
116. Kaminsky T, Gentile AM. Joint control strategies and hand trajectories in multijoint movements. *J Motor Beh.* 1986;18:261–278.
117. Marteniuk RB, MacKenzie CL, Baba DM. Bimanual movement control: information processing and interaction effects. *Q J Exp Psychol.* 1984;36A:335–365.
118. Castiello U, Bennett KMB, Stelmach GE. The bilateral reach to grasp movement. *Beh Brain Res.* 1993;56:43–57.
119. Johansson RS, Westling G. Programmed and triggered actions to rapid load changes during precision grip. *Exp Brain Res.* 1988;71:72–86.
120. Lemon RN, Mantel GWH, Rea PA. Recording and identification of single motor units in the free to move primate hand. *Exp Brain Res.* 1990;81:95–106.
121. Lemon RN, Bennett KM, Werner W. The cortico-motor substrate for skilled movements of the primate hand. In Requin J, Stelmach GE, eds. *Tutorials in Motor Behavior.* Dordrecht, The Netherlands: Kluwer; 1991:477–495.
122. Spoor C. Balancing a force on the finger tip of a two-dimensional finger without intrinsic muscles. *J Biomech.* 1983;16:497–504.
123. Chao EYS, An K-N, Cooney WP. *Biomechanics of the Hand.* Singapore: World Scientific Publishing;1989.
124. Bendz P. The functional significance of the fifth metacarpus and hypothenar in two useful grips of the hand. *Am J Phys Med Rehabil.* 1993;72:210–213.
125. Green PH. Why is it easy to control your arms? *J Motor Behav.* 1982;14:260–286.
126. Arbib MA, Iberall T, Lyons D. Coordinated control programs for control of the hands. In: Goodwin AW, Darian-Smith I, eds. *Hand Function and the Neocortex. Exp Brain Res* (Suppl 10). Berlin, Germany: Springer-Verlag; 1985:111–129.
127. Johansson RS, Westling G. Roles of glabrous skin receptors and sensorimotor memory in automatic control of precision grip when lifting rougher or more slippery objects. *Exp Brain Res.* 1984;56:550–564.
128. Johansson RS, Westling G. Tactile afferent signals in the control of precision grip. In: Jeannerod M, ed. *Attention and Performance.* Hillsdale, NJ: Erlbaum; 1990:677–713.
129. Westling G, Johansson RS. Responses in glabrous skin mechanoreceptors during precision grip in humans. *Exp Brain Res.* 1984;66:128–140.
130. Flanagan JR, Wing JM. Modulation of grip force with load force during point-to-point arm movements. *Exp Brain Res.* 1993;95:131–143.
131. Bohannon RW, Larkin PA, Smith MB, et al. Shoulder pain in hemiplegia: statistical relationship with five variables. *Arch Phys Med Rehabil.* 1986;67:514–516.
132. Wanklyn P, Forster A, Young J. Hemiplegic shoulder pain (HSP): natural history and investigation of associated features. *Disabil Rehabil.* 1996;18:497–501.
133. Wade DT, Langton Hewer R, Wood VA, et al. The hemiplegic arm after stroke: measurement and recovery. *J Neurol Neurosurg Psychiatry.* 1983;46:521–524.
134. Nakayama H, Jorgensen HS, Raaschou HD, et al. Recovery of upper extremity function in stroke patients: the Copenhagen stroke study. *Arch Phys Med Rehabil.* 1994;75:394–398.
135. Edwards S, ed. *Neurological Physiotherapy A Problem-Solving Approach.* London: Churchill Livingstone; 1996.

136. Dean C, Mackey F. Motor assessment scale scores as a measure of rehabilitation outcome following stroke. *Aust J Physiother.* 1992;38:31–35.
137. Mudie MH, Matyas TA. Upper extremity retraining following stroke: effects of bilateral practice. *J Neurol Rehabil.* 1996;10:167–184.
138. Butefisch C, Hummelsheim H, Denzler P, et al. Repetitive training of isolated movements improves the outcome of motor rehabilitation of the centrally paretic hand. *J Neurol Sci.* 1995;130:59–68.
139. Goldie P, Matyas T, Kinsella G. Movement rehabilitation following stroke. *Research Report to the Department of Health, Housing and Community Services*, Victoria, Australia; 1992.
140. Faghri PD, Rogers MM, Glaser RM, et al. The effects of functional stimulation on shoulder subluxation, arm function recovery, and shoulder pain in hemiplegic stroke patients. *Arch Phys Med Rehabil.* 1994;75:73–79.
141. Kralj A, Acimovic R, Stanic U. Enhancement of hemiplegic patient rehabilitation by means of functional electrical stimulation. *Prosthet Orthot Int.* 1993;17:107–114.
142. Pandyan AD, Power J, Futter C. Effects of electrical stimulation on the wrist of hemiplegic subjects. *Physiother.* 1996;82:184–188.
143. Hummelsheim H, Maier-Loth ML, Eickhof C. The functional value of electrical muscle stimulation for the rehabilitation of the hand in stroke patients. *Scand J Rehabil.* 1997;29:3–10.
144. Wolf SL, Catlin PA, Blanton S, et al. Overcoming limitations in elbow movement in the presence of antagonist hyperactivity. *Phys Ther.* 1994;74:826–835.
145. Krichevets AN, Sirokina EB, Yevsevicheva IV, et al. Computer games as a means of movement rehabilitation. *Disabil Rehabil.* 1995;17:100–105.
146. Shepherd RB, Carr JH. The shoulder in stroke: preserving musculoskeletal integrity for function. *Top Stroke Rehabil.* 1998;4:35–53.
147. Morris DM, Crago JE, DeLuca SC, et al. Constraint-induced movement therapy for motor recovery after stroke. *Neurorehabil.* 1997;9:29–43.
148. Carr JH, Shepherd RB. *Physiotherapy in Disorder of the Brain.* Oxford, England: Butterworth Heinemann; 1980:71–93.
149. Fitts PM. Perceptual-motor skill learning. In: Melton AW, ed. *Categories of Human Learning.* New York: Academic Press; 1964:206–284.
150. Salmoni AW. Motor skill learning. In: Holding DH, ed. *Human Skills.* 2nd ed. New York: John Wiley; 1989:197–227.
151. Fitts PM, Posner MI. *Human Performance.* Belmont, CA: Brooks/Cole; 1967.
152. Leont'ev AN, Zaporozhets AV. *Rehabilitation of Hand Function.* London: Pergamon Press; 1960.
153. van der Weel FR, van der Meer AL, Lee DN. Effect of task on movement control in cerebral palsy: implications for assessment and therapy. *Dev Med Child Neurol.* 1991;33:419–426.
154. Mathiowetz V, Wade MG. Task constraints and functional motor performance of individuals with and without multiple sclerosis. *Ecolog Psych.* 1995;7:99–123.
155. Carroll WR, Bandura A. The role of visual monitoring in observational learning of action patterns: making the unobservable observable. *J Mot Behav.* 1982;14:153–167.
156. Keele SW, Summers JJ. The structure of motor programs. In: Stelmach GE, ed. *Motor Control: Issues and Trends.* New York: Academic Press:109–140.
157. Greenwald AG, Albert SM. Observational learning: a technique for elucidating S-R mediation processes. *J Exper Psychol.* 1968;76:267–272.
158. Winstein CJ, Schmidt RA. Sensorimotor feedback. In: Holding DH, ed. *Human Skills.* Chichester, England: John Wiley and Sons; 1989:17–47.
159. Annett J, Kay H. Knowledge of results and skilled performance. *Occup Psychol.* 1957;31:69–79.
160. Newell KM. Knowledge of results and motor learning. *Exerc Sport Sci Rev.* 1976;4:195–227.
161. Thorndike EL. The law of effect. *Am J Psychol.* 1927;39:212–222.

162. Trowbridge EL, Cason H. An experimental study of Thorndike's theory of learning. *J Gen Psychol.* 1932;7:245–258.
163. Mulder T, Hulstijn W. From movement to action: the learning of motor control following brain damage. In: Meijer OG, Roth K, eds. *Complex Movement Behaviour: The Motor Action Controversy.* New York: Elsevier Science; 1988:247–259.
164. Annett J. Learning a pressure under conditions of immediate and delayed knowledge of results. *Q J Exp Psych.* 1959;11:3–15.
165. Holding DH. Learning without error. In: Smith LE, ed. *Psychology of Motor Learning.* Chicago: Athletic Institute; 1970:59–74.
166. Newell KM. Skill learning. In: Holding DH, ed. *Human Skills.* New York: John Wiley & Sons; 1981:203–226.
167. Johnson P. The acquisition of skill. In: Smyth MM, Wing AM, eds. *The Psychology of Human Movement.* London: Academic Press; 1984:215–239.
168. Carter P, Edwards S. General principles of treatment. In: Edwards S, ed. *Neurological Physiotherapy a Problem-solving Approach.* London: Churchill Livingstone; 1996:87–113.
169. Johnson P. The functional equivalence of imagery and movement. *Q J Exp Psychol.* 1982;134A:349–365.
170. Yue G, Cole KJ. Strength increases from the motor program: comparison of training maximal voluntary and imagined muscle contractions. *J Neurophysiol.* 1992;67:1114–1123.
171. Decety J, Perani D, Jeannerod M, et al. Mapping motor representations with positron emission tomography. *Nature.* 1994;371:600–602.
172. Asanuma H, Keller A. Neuronal mechanisms of motor learning in mammals. *Neuroreport.* 1991;2:217–224.
173. Crossman ERFW. A theory of the acquisition of speed-skill. *Ergonomics.* 1959;2:153–166.
174. Beggs WDA, Howarth CI. The movement of the hand towards a target. *Q J Exp Psychol.* 1972;24:448–453.
175. Canning C. Training standing up following stroke—a clinical trial. In: *Proceedings of the World Confederation of Physical Therapy Congress.* Sydney, Australia: 1987:915–919.
176. Whiting HTA. Dimensions of control in motor learning. In: Stelmach GE, Requin J, eds. *Tutorials in Motor Behavior.* New York: North-Holland; 1980:537–550.
177. Schmidt RA. *Motor Control and Learning.* 2nd ed. Champaign, IL: Human Kinetics Publishers; 1988.
178. van Rossum JHA. Schmidt's schema theory: the empirical base of the variability of practice hypothesis. A critical analysis. *Hum Movt Sci.* 1990;9:387–435.
179. Rooney KJ, Herbert RD, Balnave RJ. Fatigue contributes to the strength training stimulus. *Med Sci Sports Exerc.* 1994;26:1160–1164.
180. Marteniuk RG. Motor skill performance and learning: considerations for rehabilitation. *Physiother Can.* 1979;31:108–202.
181. Cochran BJ. Effects of physical fatigue on learning to perform a novel motor task. *Res Q.* 1975;46:243–249.
182. Brinkman JR, Hoskins TA. Physical conditioning and altered self-concept in rehabilitated patients. *Phys Ther.* 1979;59:859–865.
183. Jankowski LW, Sullivan SJ. Aerobic and neuromuscular training: effect on the capacity, efficiency and fatigability of patients with traumatic brain injuries. *Arch Phys Med Rehabil.* 1990;71:500–504.
184. Becker E, Bar-Or L, Mendelson L, et al. Pulmonary responses to exercise of patients following craniocerebral injury. *Scand J Rehabil Med.* 1978;10:47–50.
185. Hunter M, Tomberlin JA, Kirkikis C, et al. Progressive exercise testing in closed head-injury subjects: comparison of exercise apparatus in assessment of a physical conditioning program. *Phys Ther.* 1990;70:363–371.

186. Fugl-Meyer AR, Jaasko L, Leyman I, et al. The post-stroke hemiplegic patient. 1. A method of evaluation of physical performance. *Scand J Rehabil Med.* 1975;7:13–31.
187. Olney SJ, Monga TN, Costigan PA. Mechanical energy of walking of stroke patients. *Arch Phys Med Rehabil.* 1986;67:92–98.
188. Winter DA. Biomechanical motor patterns in normal walking. *J Motor Beh.* 1983;15:302–330.
189. Olney SJ, Jackson VG, George SR. Gait re-education guidelines for stroke patients with hemiplegia using mechanical energy and power analyses. *Physiother Can.* 1988;40:242–248.
190. Winter DA. *The Biomechanics of Motor Control of Human Gait.* Waterloo, Canada: University of Waterloo Press; 1987.
191. Gottlieb GL, Corcos DM, Jaric S, et al. Practice improves even the simplest movements. *Exp Brain Res.* 1988;73:436–440.
192. Hilgard ER, Bower GH. *Theories of Learning.* New York: Meredith Publishers; 1966.
193. Oxendine A. *Psychology of Motor Learning.* 2nd ed. Englewood Cliffs, NJ: Prentice-Hall; 1984.
194. Carr JH, Shepherd RB, Nordholm L, et al. Investigation of a new motor assessment scale for stroke patients. *Phys Ther.* 1985;65:175–180.
195. Cornall C. Self-propelling wheelchairs: the effects on spasticity in hemiplegic patients. *Physiother Theory Pract.* 1991;7:13–21.
196. Taub E, Wolf SL. Constraint induced movement techniques to facilitate upper extremity use in stroke patients. *Top Stroke Rehabil.* 1997;3:38–61.
197. Walsh R, Greenough W. *Environments as Therapy for Brain Dysfunction.* New York: Plenum Press; 1976.
198. Keith RA. Activity patterns of a stroke rehabilitation unit. *Soc Sci Med.* 1980;14A:575–580.
199. Keith RA, Cowell KS. Time use of stroke patients in three rehabilitation hospitals. *Soc Sci Med.* 1987;24:529–533.
200. Lincoln NB, Gamlen R, Thomason H. Behavioural mapping of patients on a stroke unit. *Int Dis Studies.* 1989;11:149–154.
201. Tinson DJ. How stroke patients spend their days. *Int Dis Studies.* 1989;11:45–49.
202. Mackey F, Ada L, Heard R, et al. Stroke rehabilitation: are highly structured units more conducive to physical activity than less structured units? *Arch Phys Med Rehabil.* 1996;77:1066–1070.
203. Esmonde T, McGinley J, Wittwer J, et al. Stroke rehabilitation: patient activity during non-therapy time. *Aust J Physiother.* 1997;43:43–51.
204. Ittelson WH, Proshansky HM, Rivlin LG, Winkel GH. *An Introduction to Environmental Psychology.* New York: Holt, Rinehart & Winston; 1974.
205. Barker RG, Wright HF. *One Boy's Day.* New York: Harper & Row; 1951.
206. Willems HF. The interface of the hospital environment and patient behavior. *Arch Phys Med Rehabil.* 1972;53:115–122.
207. Hall EJ, Flament D, Fraser C, et al. Non-invasive brain stimulation reveals reorganised cortical outputs in amputees. *Neurosci Letters.* 1990;116:379–386.
208. Cohen LG, Roth BJ, Wasserman EM, et al. Magnetic stimulation of the human cerebral cortex, an indicator of reorganisation in motor pathways in certain pathological conditions. *J Clin Neurophysiol.* 1991;8:56–65.
209. Fuhr P, Cohen LG, Dang N, et al. Physiological analysis of motor reorganisation following lower limb amputation. *Electroenceph Clin Neurophysiol.* 1992;85:53–60.
210. Jenkins WM, Merzenich MM, Ochs MT, et al. Functional reorganisation of primary somatosensory cortex in adult owl monkeys after behaviorally controlled tactile stimulation. *J Neurophysiol.* 1990;63:82–104.
211. Pascual-Leone A, Torres F. Plasticity of the sensorimotor cortex representation of the reading finger in Braille readers. *Brain.* 1993;116:39–52.

212. Pascual-Leone A, Dang N, Chen LG, et al. Modulation of muscle responses evoked by transcranial magnetic stimulation during the acquisition of new fine motor skills. *J Neurophysiol.* 1995;74:1037–1045.

213. Niemann J, Winker T, Gerling J, et al. Changes in slow cortical negative DC-potentials during the acquisition of a complex finger motor task. *Exp Brain Res.* 1991;85:417–422.

214. Seitz RJ, Roland E, Bohm C, et al. Motor learning in man: a positron emission tomographic study. *Neuroreport.* 1990;1:57–60.

215. Elbert T, Pantev C, Weinbrunch C, et al. Increased cortical representation of the fingers of the left hand in string players. *Science.* 1995;270:305–307.

216. Liepert J, Tegenthoff M, Malin JP. Changes of cortical motor area size during immobilization. *Electroenceph Clin Neurophysiol.* 1995;97:382–386.

217. Burke D. Spasticity as an adaptation to pyramidal tract injury. In: Waxman SG, ed. *Advances in Neurology, 47: Functional Recovery in Neurological Disease.* New York: Raven Press; 1988:401–423.

218. Walshe FMR. Contribution of John Hughlings Jackson to neurology. *Arch Neurol.* 1961;5:119–131.

219. Landau WM. Spasticity: what is it? What is it not? In: Feldman RG, Young RR, Koella WP, eds. *Spasticity: Disordered Motor Control.* Chicago: Year Book Publishers; 1980:17–24.

220. Landau WM. Parables of palsy, pills and PT pedagogy: a spastic dialectic. *Neurol.* 1988;38:1496–1499.

221. Tang A, Rymer WZ. Abnormal force-EMG relations in paretic limbs of hemiparetic human subjects. *J Neurol Neurosurg Psychiatry.* 1981;44:690–698.

222. Bourbonnais D, Vanden Noven S. Weakness in patients with hemiparesis. *Am J Occup Ther.* 1989; 43:313–319.

223. Dietz V, Ketelson UP, Berger SC, et al. Motor unit involvement in spastic paresis: relationship between leg muscle activation and histochemistry. *J Neurol Sci.* 1986;75:89–103.

224. McComas AJ, Sica REP, Upton ARM, et al. Functional changes in motoneurons of hemiparetic patients. *J Neurol Neurosurg Psychiatry.* 1973;36:183–193.

225. Cruz-Martinez A. Electrophysiological study in hemiparetic subjects: electromyography, motor conduction and response to repetitive nerve stimulation. *Electroenceph Clin Neurophysiol.* 1984;23:139–148.

226. Farmer SF, Swash M, Ingram DA, et al. Changes in motor unit synchronization following central nervous lesions in man. *J Physiol (Lond).* 1993;463:83–105.

227. Tsuji I, Nakamura R. The altered time course of tension development during the initiation of fast movement in hemiplegic patients. *Tohoku J Exp Med.*1987;151:137–143.

228. Knutsson E, Martensson A. Dynamic motor capacity in spastic paresis and its relation to prime mover dysfunction, spastic reflexes and antagonist coactivation. *Scand J Rehabil.* 1980;12:93–106.

229. Giuliani CA. Adult hemiplegic gait. In: Smidt GL, ed. *Gait in Rehabilitation.* New York: Churchill Livingstone; 1990:253–266.

230. O'Dwyer NJ, Ada L, Neilson PD. Spasticity and muscle contracture following stroke. *Brain.* 1996; 119:1737–1749.

231. Darian-Smith I, Galea MP, Darian-Smith C. Manual dexterity: how does the cerebral cortex contribute? *Clin Exp Pharmacol Physiol.* 1996;23:948–956.

232. Lance JW. Symposium synopsis. In: Feldman RG, Young RR, Koella WP, eds. *Spasticity Disordered Motor Control.* Chicago: Year Book Medical Publications; 1980:485–494.

233. Lance JW. What is spasticity? [Letter] *Lancet.* 1990;335:606.

234. Ashworth B. Preliminary trial of carisoprodol in multiple sclerosis. *Practitioner.* 1964;192:540–542.

235. Wartenburg R. Pendulousness of the legs as a diagnostic test. *Neurol.* 1951;1:18–24.

236. Bohannon RW, Smith MB. Interrater reliability of a modified Ashworth Scale of muscle spasticity. *Phys Ther.* 1987;67:206–207.

237. Fowler V, Canning CG, Carr JH, et al. The effect of muscle length on the pendulum test. *Arch Phys Med Rehabil.* 1997;79:169–171.

238. Katz RT, Rymer WZ. Spastic hypertonus mechanisms and measurement. *Arch Phys Med Rehabil.* 1989;70:144–155.

239. Dietz V, Quintern J, Berger W. Electrophysiological studies of gait in spasticity and rigidity: evidence that altered mechanical properties of muscle contribute to hypertonus. *Brain.* 1981;104:431–449.

240. Hufschmidt A, Mauritz K-H. Chronic transformation of muscle in spasticity: a peripheral contribution to increased tone. *J Neurol Neurosurg Psychiatry.* 1985;48:676–685.

241. Thilmann AF, Fellows SJ, Rose HF. Biomechanical changes at the ankle joint after stroke. *J Neurol Neurosurg Psychiatry.* 1991;54:134–139.

242. Carey JR, Burghardt TP. Movement dysfunction following central nervous lesions: a problem of neurologic or muscular impairment. *Phys Ther.* 1993;73:538–547.

243. Dietz V, Trippel M, Berger W. Reflex activity and muscle tone during elbow movements in patients with spastic paresis. *Ann Neurol.* 1991;30:767–779.

244. Williams PE, Goldspink G. Changes in sarcomere length and physiological properties in immobilized muscle. *J Anat.* 1978;127:459–468.

245. Gossman MR, Sahrmann SA, Rose SJ. Review of length-associated changes in muscle. *Phys Ther.* 1982;62:1799–1808.

246. Herbert R. The passive mechanical properties of muscle and their adaptations to altered pattern of use. *Aust J Physiother.* 1988;34:141–149.

247. Williams RG. Sensitivity changes shown by spindle receptors in chronically immobilized skeletal muscle. In: *Proceedings of the Physiological Society.* 1980:26P–27P.

248. Gioux M, Petit J. Effects of immobilizing the cat peroneus longus muscle on the activity of its own spindles. *J Appl Physiol.* 1993;75:2629–2635.

249. Proske U, Morgan DL, Gregory J. Thixotropy in skeletal muscle and in muscle spindles: a review. *Prog in Neurobiol.* 1993;41:705–721.

250. Halligan PW, Marshall JC. Left visuo-spatial neglect: a meaningless entity? *Cortex.* 1992;28:525–535.

251. Riddoch MJ, Humphreys GW. Towards an understanding of neglect. In: Riddoch MJ, Humphreys GW, eds. *Cognitive Neuropsychology and Cognitive Rehabilitation.* London: Erlbaum; 1994:125–149.

252. Ladavas E, Petronio A, Umilta C. The deployment of visual attention in the intact field of hemineglect patients. *Cortex.* 1990;26:307–317.

253. Maugiere F, Desmedt JE, Courjon J. Astereognosis and dissociated loss of frontal or parietal components of somatosensory evoked potentials in hemispheric lesions. *Brain.* 1983;106:271–311.

254. Calvanio R, Levine D, Petrone P. Elements of cognitive rehabilitation after right hemisphere stroke. *Behav Neurol.* 1993;11:25–57.

255. Geschwind N. The apraxias: neural mechanisms of disorders of learned movement. *Am Scientist.* 1975;63:188–195.

256. Wade DT. *Measurement in Neurological Rehabilitation.* Oxford, England: Oxford University Press; 1992.

Chapter 3

Skill Acquisition: Action, Movement, and Neuromotor Processes

A. M. Gentile

Rehabilitation therapists teach functional skills. They try to help patients regain mastery in the performance of everyday tasks. To help patients learn or relearn functional activities, therapists must first understand how skill is acquired. The purpose of this chapter is to examine those processes underlying skill learning and then to explore implications for therapeutic practice.

The chapter is organized into four major sections. The foundation for later discussion is set in the first section. Some common terms are defined and basic concepts are examined. Different types of goal-directed behaviors are described, with functional activities identified as most relevant for therapists. These functional behaviors are analyzed on three levels: (1) action; (2) movement; and (3) neuromotor processes. The relationship between these levels is discussed, leading to a definition of skill.

In the second section, a taxonomy of tasks is presented. Tasks are categorized based on the environmental context in which an action takes place and the action's functional role. Classifying tasks in this way enables therapists to identify the information-processing and motoric demands placed on the patient, and provides insight into those processes underlying skill learning.

Acquisition of skill is discussed in the third section of this chapter. Skill acquisition is proposed to be mediated by two interdependent learning processes.[1] One process is consciously available to the performer and, thus, is referred to as explicit learning. The second process involves changes in performance that are not consciously accessible and is referred to as implicit learning. How these two processes influence the development of skill is analyzed during initial and later phases of learning. Analyzing tasks and learning processes provides the conceptual framework for the final section of the chapter. It is at this point that the question of most interest to therapists is addressed: How do we intervene in these processes to help patients acquire skill?

BASIC CONCEPTS

Goal-Directed Behavior

The behavior that dominates our daily lives is directed toward the accomplishment of goals.[2,3] Such behavior is aimed at a specific purpose or end that we are trying to achieve. It is intentional; it is linked to outcomes we are attempting to produce. It has the quality of perseverance. The behavior continues in time until the goal is reached (or until we give up and change the task). Goal-directed behavior is guided by the consequences it produces—guided by feedback that informs us as to how close or how far we are from accomplishing our objective.

We can distinguish between several types of goal-directed behaviors:

1. Investigatory behaviors
2. Adaptive behaviors
   a. Functional behaviors: Interactions with the physical environment
   b. Communicative behaviors: Interaction with the social environment

Investigatory Behaviors

Investigatory behaviors are used to gather information from the environment. For example, as we move our hand over an object's surface, we obtain information about its texture, temperature, or shape. In a similar fashion, we use head and eye movements to visually search and pick up information from our surroundings. To localize the source of a sound, the head is moved sideways so that sound pressure patterns arriving at the two ears differ in time. We use mouth and tongue movements to enhance gustatory input and we sniff to amplify olfactory cues. In all of these examples, the behaviors are not directed toward changing the environment. Rather, these investigatory behaviors are used to orient, focus, or adjust our sensory analyzers to best gather information. Although investigatory behaviors are not the primary concern of this chapter, they are discussed with reference to their role in providing critical information about the environmental context.

Functional Behaviors

Functional behaviors enable us to interact with the physical environment. In general, these behaviors involve changing or maintaining body orientation, changing or maintaining the position of objects, or doing both concurrently. Throughout the day, we set goals to maintain body position (remain seated in a chair; stand waiting in a line) or change body position (move about in our surroundings; rise up from a chair; lie down on a couch). Similarly, we seek to maintain the position of objects (carry books; balance a cup) or change their positions (pick up a package; throw a ball; push a tray forward). Often, we set functional goals that concurrently involve body-orientation and interactions with objects (walk forward to pick up an object; carry a child while crossing a busy street; sit holding a newspaper so as to read). These functional behaviors enable us to cope with our everyday physical surroundings and are of most interest to rehabilitation therapists.

Communicative Behaviors

The second type of adaptive behavior involves interaction with the social environment. We use gesture, "body language," facial expression, sound production, and speech to interact with others. The purpose of these behaviors is the transmission of information from one person to another. Motor control processes underlying speech and limb movements share many common features.[4,5] However, this chapter focuses on functional actions. Thus, communicative behaviors are not discussed further.

Levels of Analysis: Action, Movement, and Neuromotor Processes

Functional behaviors can be analyzed on three levels: (1) actions; (2) movements; and (3) neuromotor processes. Let us consider the defining characteristics associated with each level.

Action

Action is the observable outcome resulting from the performer's purposeful interaction with the environment. The performer's internal representation of an intended outcome is the action-goal. Examples may help to clarify this concept. Suppose, at one point in time, you observe the relationship between the performer and the environment (initial conditions). Action is the change from those initial conditions to some later state. For instance, you are with a patient who is wearing pajamas and has decided to change clothes (action-goal). You are called away from the room. When you return, the patient is now dressed in street clothes. Action is the observable change from the initial to final conditions; it is the outcome resulting from the performer-environment interaction. Another example: a mother speaking to a child says, "When I come home I want to see this room cleaned." The mother is demanding action. She is specifying the outcome to be produced. In one last example, you give your handwritten report to the secretary to type. Returning some time later, you observe that the report has been typed (action has occurred) or has not been typed (no action has taken place).

Actions are not always successful. A change in state may occur without the action-goal being reached. Using the previous example in which a patient has set a goal of dressing in street clothes, you may observe that action has occurred (the patient is no longer wearing pajamas). However, the outcome is not as intended (shirt buttons are open; jacket is only halfway on). Similarly, your secretary may give you the typed report replete with spelling errors and a major section missing. She has acted but none too successfully.

To infer that action has taken place, it is not necessary to observe the performer's movement. In the example given previously, no one observed how the patient managed to dress in street clothes. Further, the mother is not indicating how the child should clean the room. Finally, you do not have to watch the typing movements of your secretary to judge whether he or she has acted on your report. Action is defined by the end-state or outcome resulting from the performer-environment interaction and does not implicate how that end is achieved.

Movement

Movement is the means through which action-goals are accomplished. To successfully produce an intended outcome, movements must exactly match certain features of the physical environment. The action-goal determines which features of the environment are critical. For example, to successfully climb stairs, the foot must be lifted to a height not less than that of the step; it must be positioned in accord with the step's width; and the texture of the step's surface must be considered as force is applied against it. As another illustration, think of reaching forward to grasp a cup. Hand and fingers must shape exactly to the cup's configuration. In both these instances, the performer is not free to decide on the movement's spatial form. Rather, critical features of the environment determine the movement's spatial arrangement if the action-goal is to be reached. Sometimes, the movement's timing also is constrained. Imagine a child trying to catch a ball. Not only must the child shape his or her grasp to the ball's spatial dimensions, but timing the grasp motion and contact with the ball is also controlled by the ball's flight. Thus, to successfully attain an action-goal, movement patterns may have to match spatial features alone or may have to match both spatial and temporal features of the environment.

Regulatory conditions are those environmental features to which the movement must mold to successfully reach the action-goal.[6] Background features of the environment, which are irrelevant for movement organization, are considered *nonregulatory*.[6] Suppose the action-goal is to throw a dart at a target. Regulatory conditions are the ones to which the movement must conform if the goal is to be achieved. Hence, movement organization is critically determined by the dart's size, shape, and weight and the target's distance, height, and size. In contrast, the dart's color does not specify any aspect of the throwing movement and, therefore, is nonregulatory. If an audience is watching this dart-throwing event, people's cheers or negative comments may affect the motivational state of the performer. However, the audience's reactions are nonregulatory as they do not specify how the movement is to be organized to attain the goal.

As suggested by Bernstein,[7] a movement pattern can be described in terms of its overall form (topology of the movement) or in terms of specific spatial/temporal parameters (metrics of the movement). What is meant by topology is the general, configurational aspects that remain relatively stable although the movement may be scaled up or down in amplitude, timing, or force. Topology can be thought of as an envelope defining the direction and flow of movement, extended in time, but without detailed patterning. For example, the movement used to propel a dart to a target can be described qualitatively as "throwing" (its topology). In contrast, the spatial/temporal parameters of a multi-joint arm motion can be measured precisely in terms of displacement, velocity, and acceleration (its metrics). The movement's topology and metrics emerge from a mapping between regulatory environmental features and the performer's body morphology. The interplay between these two structured entities—the person and the environment—determines the movement's general form and detailed patterning.

Neuromotor Processes

Neuromotor processes are organized in advance of the observable movement. In other words, movement is directed by a plan. How motor plans are organized is not well understood. We do have some insights, however, into the general nature of motor control that help us think about this planning process.

First, neural processes associated with motor planning are not restricted to one site within the central nervous system (CNS). There is no strict localization of a motor memory trace; no cluster of executive neurons specifies all aspects of the movement pattern.[7] Further, activities at one site within the CNS are not isomorphic to the movement; these activities do not represent the patterning of movement in all of its detail. Rather, neuromotor control is distributed over several subsystems, located at different sites within the CNS.[8,9] Activity within a subsystem is directed toward solution of a particular problem in the movement's organization. Each subsystem has access to specific forms of information. One subsystem may access detailed spatial information about the external environment representing an integration of visual, auditory, and somatosensory inputs. Another subsystem may have immediate access to inputs from cutaneous, joint, and muscle receptors specifying status of the periphery (musculoskeletal structures). Still other subsystems may receive a combination of proprioceptive and other afference necessary for intersegmental control processes. In addition, a continuous flow of information exists among these various subsystems. Communication is provided by a network of rich interconnections. The order of communication within this network does not constitute a fixed hierarchy in which one subsystem always sends instructions to be carried out by another. Rather, the focus of activity shifts according to which subsystem has most immediate access to information relevant for the behavioral task. Although focus may shift, activity within each subsystem is ongoing and operates in parallel with that of others.

The modes of interaction across these various subsystems can be viewed metaphorically as an overlaying of dispositions to move. Within the ongoing interaction, one subsystem might influence the movement's general form; another subsystem may constrain the movement's spatial and temporal parameters; still others may provide rules for coordination. The patterning of movement we observe is an emergent form, resulting not from activity within one subsystem but from the reciprocity of influence and consensus of activity across subsystems.[7]

Relationships Between Action, Movement, and Neuromotor Processes

The relationship between these three levels of analysis is not one-to-one. Instead, many movements can be used to achieve an action-goal. Similarly, neuromotor processes can be organized in many ways to have a particular movement emerge. As shown diagrammatically in Figure 3–1, the relationship between these levels can be characterized as many-to-one. This relationship between movement patterns and the action-goal is referred to as movement equivalence. As an example, consider the many movements you could use to throw an object at a target (eg, underhand, overhand, or sidearm). Any one of these movement forms could produce the same outcome (ie, achieve the action-goal by hitting the target).

In a similar fashion, one particular movement pattern is not reducible to a fixed mode of organization within the CNS. Neuromotor processes are organized dynamically, in flexible ways, to yield the same movement pattern. This many-to-one relationship between neuromotor processes and movement is referred to as motor equivalence.[10]

SKILL

The lack of a fixed relationship between action, movement, and neuromotor processes is important for our concept of skill. Skill is not determined by using one specific movement pattern or by organizing neuromotor processes in one set way. Rather, skill is defined as consistently attaining an action-goal with some economy of effort. It involves an individual solution to the problem of how to efficiently organize movement to produce an action-outcome consistently.

Movement patterns are derived, in part, from human ingenuity. They represent a unique match between two structured entities: the performer and the

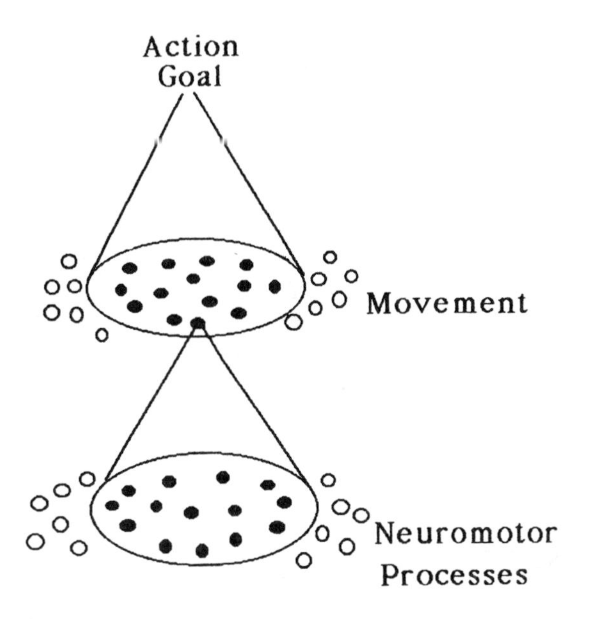

Figure 3–1 Diagrammatic representation of the relationship between levels of analysis. At the "Movement" level, filled circles represent the many movement patterns that can be used to successfully achieve the "Action-Goal" (movement equivalence); unfilled circles are unsuccessful patterns. At the level of "Neuromotor Processes," filled circles represent the many ways neural processes can be organized to produce a specific movement (motor equivalence); unfilled circles are unsuccessful modes of organization.

---

environment. Performers are structured in different ways. They vary in height, weight, limb length, strength, endurance, and flexibility. Structure of the physical environment also varies. Common objects, such as chairs, differ along many dimensions: height from the floor, width of the seat, solidity, texture of the surface, absence or presence of arms, even general configuration. No single movement pattern can be used successfully by all people for sitting on a specific chair. Further, a person cannot use one fixed movement pattern to sit on different chairs. Shaped by the performer's dimensions and the environment's constraints, skill involves movement patterns that consistently and efficiently achieve the action-goal.

In trying to help patients regain functional skills, rehabilitation therapists set goals for the patient and arrange the environment in which the action takes place. However, it is the patient who must organize a movement that matches the environment and produces the desired outcome. For example, imagine that the task set for a stroke patient is to climb a set of three wooden stairs. The therapist may intervene in this situation, suggesting how the movement might be organized. However, it is the patient that must solve the problem of how to compose and control a movement that will transport the body from the starting position to the top of the stairs.

Skill is task-specific. The movement pattern used to successfully climb these three wooden stairs cannot be used to climb carpeted stairs or stairs of a different height. The task has changed and so have the demands placed on the performer. As the task-structure changes, movement must be reshaped to fit new constraints posed by the environment.

Different tasks pose different requirements on the performer. Information-processing demands vary. So do the organizational constraints placed on neuromotor processes. Therapists need to understand the task-demands placed on the patient. To facilitate skill learning, therapists must be able to analyze the precise nature of tasks to devise effective therapeutic interventions.

ANALYSIS OF TASKS

In this section, a system for analyzing functional activities is presented: a taxonomy of tasks. The taxonomy provides a framework for understanding the requirements placed on a performer and the processes involved in skill acquisition.[6,11] Tasks are analyzed from two perspectives. First, we examine the environmental context in which the action takes place. Second, we determine the action's functional role. Analyzing tasks from these two perspectives yields a taxonomy consisting of 16 task-categories.

Environmental Context

Let us start by analyzing tasks according to the environmental context in which the action takes place. Recall that movement must be shaped to match critical features of the environment to achieve the action-goal. It must conform to the regulatory environmental conditions to successfully carry out a task. In evaluating the environmental context, tasks are categorized: (a) by specifying the regulatory conditions during performance; and (b) by determining whether these conditions change from one attempt to the next.

Regulatory Conditions during Performance

Two types of tasks can be distinguished by examining the environmental context during a performance. First, there are environments in which objects or other people are stationary and the terrain is fixed. Second, there are environments in which the supporting surfaces, objects, or people are in motion.

Stationary Environments. When regulatory conditions involve a fixed terrain and stationary objects, the environment controls only spatial parameters of the movement (eg, the movement's shape, size, and extent). Examples of such tasks are sitting in an armchair, walking down an empty corridor, putting on a sweater, and buttoning a shirt. In all these tasks, movement must be configured to fit spatial features of objects and supporting surfaces. Imagine picking up a newspaper. The size and shape of the paper specifies how you must spatially organize your movement. You might pick up the paper by opposing thumb and fingers, by using a precision grasp, by using two hands, or with a rather novel solution, by using opposing elbows. Although these movements vary greatly in form, all must have one feature in common: the distance between opposing body parts must match the dimensions of the paper.[12,13] Hence, spatial parameters of the movement must conform to regulatory spatial features for a successful outcome to be produced.

When regulatory conditions involve a fixed terrain and stationary objects, timing is not specified. Therefore, the movement is self-paced. The performer is free to decide when to start, when to end, how long to continue the movement, and how to temporally sequence components. Further, all information about the regulatory conditions is immediately available in the environment. To pick up relevant information, the performer can visually scan the surroundings in an unhurried fashion.

Motion in the Environment. The second type of task involves environments in which objects, other people, or supporting surfaces are in motion. For example, think of sitting on a moving train, catching a thrown ball, crossing a busy intersection, stepping onto an escalator, or picking up a squirming child. In all these situations, motion in the environment occurs independently of the performer's movement. Now, both spatial and temporal features of the environment constrain the performer's movement. Thus, movement is externally paced; the movement's timing is determined by the external environment.

When the performer's movement must conform to the motion of other objects, certain complexities are introduced. To be coincident with a moving object, the performer has to compensate for delays inherent in processing information and in executing the movement. These intrinsic time-lags are not evident when movement is carried out under stationary conditions. However, they are important considerations when we cope with an environment in motion.

Subjectively, our experience of reality is that we live in the immediate present. As we engage in conversation, we think that we hear and see the speaker at the same instant in time as the speaker is producing speech. However, several delays exist between speech production and our perception of it. It takes time for the sound pressure patterns produced by the speaker to reach our ears. Additional time is lost during the translation of this energy pattern into neural events. Then, neural transmission adds further delays. By the time we "hear" what has been said, the speaker may well be saying something else. Similar delays are involved in the processing of visual signals. Therefore, our impression of being "in the present" is somewhat in error. Actually, we live in the immediate past.

Other time-lags are introduced when we try to respond to external events. Neural processes associated with motor planning take time, as does neural transmission to muscles. Furthermore, muscles and bony structures have certain inertial characteristics that must be overcome before the movement starts. Thus, time taken for neuromuscular processes and biomechanical events adds to the delays in information-processing.

Consider a situation in which an individual is instructed to initiate a movement as soon as a light is activated. The elapsed time between the onset of the light stimulus and the initiation of the response is a measure of reaction time. If the subject cannot predict when the light will go on, reaction time typically averages about one-fourth of a second. Lagging behind the real world by less than a second does not seem very important; and it is not, as long as objects or people remain stationary. However, when the environment is in motion, these time-lags associated with processing information and executing movement can be very detrimental to performance. If you aim for where a moving object was one-fourth of a second ago, it is no longer there now. When coping with objects in motion, you cannot organize movement on the basis of information presently available. Rather, you must go beyond the information given.

When objects are in motion, the performer must extrapolate from the object's present path and velocity to determine where it will be some brief time hence. Within our visual system, several mechanisms pick up the velocity characteristics of moving objects.[14] These mechanisms assist in this extrapolation process as long as a stable trajectory is maintained. However, objects and, especially, people change paths and rates of motion. Determining where an object will be requires prediction of future events. For example, as two people walk toward each other, they are both anticipating the other's path and rate. Sometimes in crowded corridors, one steps aside to let the other pass. Sometimes, both step to the same side simultaneously because each person predicts that the other will continue to walk forward. You may have witnessed (or experienced) two people going through a series of right-left side steps and neither walking forward. The breakdown of customary performance in this situation is not due to movement deficiencies. Rather, predictive processes have led to the snag.

In coping with complex environments, where many objects and people are in motion, the performer makes ongoing predictions about the unfolding scene. These predictions compensate for inherent time-lags and provide for effective movement. To pick up advance cues from a complex visual array, the performer's investigatory movements become attuned to dynamically evolving events. Skilled performers know where to look in a complex environment to gather crucial information.

Remember the process you went through in learning to drive a car. Initially, you scanned the road immediately in front of the automobile. Then, you found that this close focus did not allow sufficient time to process information. Looking farther ahead provided more time to analyze inputs and, therefore, to control the car's motion more effectively. In addition, you began to recognize certain cues signaling upcoming events: a ball rolling into the road may be followed by a child; erratic motion of another car may foreshadow some dangerous move; or a blinking traffic light in the distance may mean adjacent cars will slow down. In picking up these advance cues, upcoming events were anticipated and movement was better organized. In tasks involving motion in the environment, such predictive processes provide the bridge to future events enabling skilled performance.

Regulatory Conditions: Intertrial Variability

Thus far, we have analyzed conditions present during performance and have identified two types of tasks (environments: stationary versus in motion). Now, the environmental context is examined from another perspective. Tasks are analyzed to determine whether the regulatory conditions remain the same or change from one performance to the next. Variation from one attempt to the next ("intertrial variability") has important implications for skill acquisition. Three factors are affected: (1) the demands placed on attentional processes; (2) the organization of movement; and (3) the mode of representation in memory.

Intertrial Variability: Absent. Practice under unchanging conditions increases the predictability of environmental events: there is no uncertainty about what conditions will prevail. Thus, information-processing is minimal. As the situation is known to the performer, the need to continuously monitor the environment is reduced. Only intermittent sampling of relevant information is required to guide performance.[15] For example, repetitive knitting can be carried out while watching television by visually switching from one activity to the other. Although we can attend to only a limited number of events, high predictability of environmental conditions saves attentional-resources and opens the range of activities that can be undertaken simultaneously.[16]

Although monitoring of the environment may decrease with practice, regulatory inputs that were present during initial learning cannot be withdrawn or changed without disrupting performance. As reviewed by Proteau and colleagues,[17,18] research contradicts prior notions that learning leads to less dependence on visual information and more reliance on kinesthetic inputs. If initial learning is carried out based on visual information from the environment, this information must be available or performance will deteriorate. Hence, learning is highly task-specific.

Intertrial variability also influences movement organization. During initial practice of any new task, the performer must devise a general movement form (topology) that shapes to the environment and must adapt the movement's temporal and spatial parameters to match task-constraints. Active exploration of alternatives is evident as the performer tries to settle into a movement organization that accomplishes the action-goal. When the environment does not change, this "settling-in" on a particular movement organization is reflected by less variability in performance.[19–23] With continued practice, the movement's topology stabilizes. As regulatory conditions are unchanging, specification of the movement's temporal and spatial parameters becomes more precise. This process of honing in on a general topology and on specific parameters is referred to as movement fixation.

As learning progresses, neuromotor processes seem to be operating in a reproductive mode. Attentional demands associated with planning the movement decrease; performance becomes habitual and automatic. Memory associated with the movement's topology is consolidated, as is specification of the movement's temporal and spatial parameters. These memory processes related to movement organization are variously referred to as schemas, motor programs, or internal models.[24–26] In the present discussion, the latter term is used because it is theoretically neutral (unlike the term "schema") and avoids the computer metaphor (ie, "programs").

When environmental conditions do not change over trials, the internal model of the movement's topology is activated, similar parameters are set, and reproduction of the movement is attempted. When the environment does not vary, these attempts at movement-reproduction become more refined and consistent with practice.[27]

Intertrial Variability: Present. When the environmental context changes from one attempt to the next, movement must be adapted to the new circumstances. To successfully attain the action-goal, there must be as many movement variations as there are changes in the environment. Varying the movement pattern may mean keeping the same topology while specifying new spatial or temporal parameters (eg, walking faster or slower, grasping a larger or smaller object, altering the timing of a forehand stroke in tennis). Sometimes, however, environmental change exceeds a critical threshold and mandates a new movement topology. For instance, when the size of an object is too large for grasping with one hand, lifting with two hands is required. Again, using tennis as an example, a change in the ball's trajectory with reference to the player's position may mandate a shift from a forehand to a backhand stroke. The goal is the same (hit the ball over the net). However, the marked change in regulatory conditions (direction of the ball's path) requires a major reorganization of the movement pattern (a new topology).

Thus, practice under variable regulatory conditions diversifies the movement's organization. This is in sharp contrast to the movement fixation observed when environmental conditions do not vary from one attempt to the next. The results of a study by Higgins and Spaeth[28] illustrate this contrast. They compared dart throwing movements under conditions of constant or variable practice. In Figure 3–2 (left side) is a representation of the movement pattern that developed with repeated practice under environmental conditions that remained constant. Note the decrease in movement variability as the performer hones in on a successful pattern. On the right side of Figure 3–2, practice under variable conditions diversifies the movement pattern to match changes in the environment. Thus, variable practice develops a flexibly organized, movement repertoire: each movement pattern shaped in whole or in part to variation in environmental events.

When regulatory conditions vary, neuromotor processes seem to operate in a generative mode. What is learned is how to compose a movement pattern "on the spot" to fit variable environmental events. The need to restructure topology or redefine parameters implicates ongoing problem solving rather than stabilizing only one mode of movement organization in memory. With practice under variable conditions, movement organization is tuned by inputs specifying status of the environment and performer. Such tuning limits the range of possible movement options. However, the problem of precisely fitting movement to environmental constraints must be solved anew for each alteration in regulatory conditions.

Unlike tasks in which the situation remains constant, ongoing monitoring of the environment is necessary to cope appropriately with variable regulatory conditions. Surveillance is essential to detect change. Allocation of attentional resources is required to process information and plan the movement. Using the distinction made by Shiffrin and Schneider,[29] performance under variable

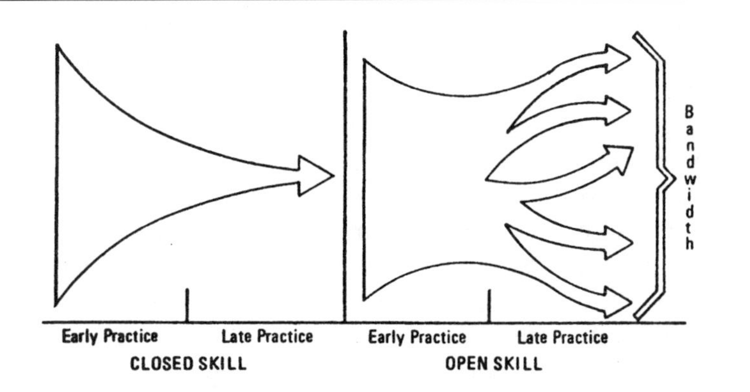

Figure 3–2 Change in variability of movement during practice of Open and Closed tasks. *Source:* Reprinted with permission from J.R. Higgins and R.K. Spaeth, Relationships Between Consistency of Movement and Environmental Condition, *Quest*, Vol. 17, pp. 61–69, © 1972, Human Kinetics.

conditions can be characterized as "controlled" in contrast to the "automatic" performance observed when conditions are constant.

Although variable environments pose high information-processing demands, the performer can simplify the task by developing probability functions as to the likelihood that particular events will occur.[30] Assigning probabilities to environmental conditions enables the advance organization of appropriate movement options (priming). Such advance preparation reduces the time required for movement organization. However, some risk is involved. Unlikely events do occur and catch the performer off guard. Thus, using probability functions in tasks with high intertrial variability must be balanced against the evaluation of risk.

The Structure of Tasks: Environmental Context

Now, we can sketch out the first dimension of the task taxonomy based on environmental context. We analyzed tasks based on regulatory conditions during performance and classified them as involving environments that are stationary or in motion. Similarly, intertrial variability can be absent or present. Combining these two ways of analyzing the environmental context yields four distinct categories: (a) Closed tasks, (b) Consistent Motion tasks, (c) Variable Motionless tasks, and (d) Open tasks. Shown in Table 3–1 are examples of tasks within each of these categories. Characteristics of these four task-categories are summarized in Table 3–2.

In *Closed* tasks, regulatory conditions involve objects, other people, or supporting surfaces that are stationary during performance and do not change over successive attempts. These tasks are designated as "Closed" because interaction between performer and environment is less than in other situations. Control is centered more with the performer.

In *Variable Motionless* tasks, all aspects of the environment are stationary but spatial features may change from one attempt to the next. For example, during each day, a performer typically walks on quite different surfaces including carpet, gravel, cement, or slick tile. In each case, the supporting surface is immobile but its texture differs markedly.

In *Consistent Motion* tasks, objects or supporting surfaces are in motion but do not change over successive attempts. In these tasks, motion is consistent because it is produced by some electrical or mechanical device. A person's motion (or motion imparted to an object by a person) is rarely the same from one attempt to the next. Human motor control is inherently variable. As an example of the natural variability in human motion, think of a father throwing a ball to his young child. The father tries to throw the same way each time so as to simplify the task. Over successive attempts, you will hear the father say, "No, that was daddy's error: I threw too high" (or too low, or too far, etc.). Achieving the same trajectory of the ball is no simple feat. Therefore, when we want

Table 3–1 Environmental Context: Task Categories and Examples

Regulatory Conditions during Performance	*Intertrial Variability*	
	Absent	*Present*
	Closed Tasks	**Variable Motionless Tasks**
Stationary	• Climbing stairs at home • Brushing teeth • Unlocking the front door • Stepping on the bathroom scale	• Walking on different surfaces • Climbing stairs of different heights • Drinking from mugs, glasses, cups
	Consistent Motion Tasks	**Open Tasks**
Motion	• Stepping onto an escalator • Lifting luggage from an airport conveyor • Moving through a revolving door	• Sitting in a moving automobile • Catching a ball • Walking down a crowded hall • Carrying a wiggling child

Source: Adapted from *Structure of Motor Tasks* (p 12) by AM Gentile et al, Professionalle de L'Activite Physique du Quebec, with permission of the author, © 1975.

Table 3–2 Environmental Context: Task Characteristics

Regulatory Conditions during Performance	*Intertrial Variability*	
	Absent	*Present*
Stationary	• Self-paced • No predictive demands • Movement fixation • Reproductive mode • Monitoring decreases	• Self-paced • No predictive demands • Movement diversifies • Generative mode • Monitoring ongoing
Motion	• Externally paced • Predictive demands • Movement fixation • Reproductive mode • Monitoring decreases	• Externally paced • Predictive demands • Movement diversifies • Generative mode • Monitoring ongoing

motion in the environment to be relatively constant, external devices are used (as shown by the examples in Table 3–1).

In *Open tasks*, objects, other people, or supporting surfaces are in motion and conditions change over trials. The performer and the environment are dynamically interacting. Tasks in this category are the most complex and place the most demands on the performer.

Artificially Imposed Timing Controls. In addition to the control normally exercised by the environment's physical features, humans have devised ways of imposing artificial timing prescriptions on actions. Sometimes, external signals specify when to go or stop or how long the action should continue. For example, red and green traffic lights regulate the starting or stopping of vehicular or pedestrian movement. As another example, a performer is sometimes instructed to carry out the task as rapidly as possible or within a prescribed time interval. External signals or instructions specifying some aspect of an action's timing do not specify how the movement is to be organized (ie, they do not determine how to move, only when or at what rate). Hence, these artificially imposed temporal controls are different from regulatory environmental conditions. Timing signals, devised by humans, are an extension of social control. Such artificial controls can be imposed on tasks within any of the four categories related to environmental context.

Functional Role

A second way in which tasks can be analyzed relates to the functional outcome guiding action. The action's functional role may require maintaining or changing body orientation, maintaining or changing the position of objects, or doing both concurrently. Processing of information differs as the action's function changes. Let us consider the characteristics of tasks directed toward different functional ends.

Body Orientation

Orientation of the body is specified by the action-goal. The task can require that a stable body position be maintained (Body Stability) or the task can necessitate transporting the body from one place to another (Body Transport). For both types of tasks, environmental information is used to specify surfaces offering support. As suggested by Gibson,[31] the performer must detect whether the terrain will accommodate standing or walking and whether surfaces will support sitting or leaning. Thus, all tasks require detecting solidity of the terrain or support-surfaces.

Body Stability. Information-processing demands are relatively low in tasks requiring maintenance of a stable body position. Boundaries of the regulatory environment are fixed to an area immediately surrounding the performer. When body position is stable, the effective boundary falls within the performer's reaching distance ("graspable space"). Events within this area affect the movement's organization; more distant events do not. For example, if the action-goal is to maintain a stable body position while sitting on a moving train, then information from the immediate area must be picked up and analyzed to organize the sitting movement effectively. In this situation, relevant information would include the seat's height, depth, solidity, texture; texture of the surface under the feet; and change in support surfaces induced by the train's motion. Obviously, other people walking in the train's aisle are irrelevant (nonregulatory) for control of this sitting task.

Body Transport. When the task requires transporting the body from one place to another, information-processing demands are high. Boundaries of the environment are constantly expanding in the direction of the performer's motion. Thus, the performer must look ahead to pick up relevant information. How far ahead to look is determined by the rate of body motion. For example, when moving forward rapidly, as in running, the performer has to visually scan the environment far ahead. Because of intrinsic time-lags, the area immediately in front of the person does not provide usable information (it is "dead-space"). By the time information is picked up and analyzed, it is too late to modify the movement. The extent of this dead space is determined by how fast the person is moving. As the rate of body transport increases, so does the region of unusable information.

In scanning the environment, the performer considers three factors: (1) rate of body motion; (2) time-lags posed by processing delays; and (3) risk-level tolerable. In many ways, body transport involves predictive processes similar to coping with moving objects. However, an important difference exists between these two types of tasks. Coping with motion of external objects is outside of the performer's control. In contrast, a person has considerable control over the situation in body transport tasks. The individual can decide to slow down the rate of forward motion. This option is particularly important when information concerning the upcoming environment is not accessible (eg, rounding a corner while walking) or when risk of serious accident is high (eg, driving an automobile in fog). Evaluating trade-offs between rate of motion and risk-level adds to the information-processing demands in body transport tasks.

Body Orientation versus Postural/Locomotor Systems. In many instances, the action's functional role and neuromotor control processes are inseparable. Body stability and body transport tasks are usually, but not always, carried out by neuromotor systems mediating postural support or locomotion. Certainly, sitting in a chair (body stability) implicates postural support systems. Similarly, transporting the body by walking implicates neuromotor processes controlling locomotion.

However, there are tasks in which the action's functional role and neuromotor control processes are different. For example, think of driving a car, riding on an escalator, or standing on one of those moving floors at an airport. In all these tasks, the functional role is body transport; however, neural processes are regulating postural support. In some instances the functional role is body stability while the neuromotor processes are engaged in a fashion typically used in locomotion. Examples of such situations are running on a treadmill or peddling an exercycle. Whenever a split exists between functional role and neuromotor control, some electrical or mechanical device is involved. Humans have ingeniously designed devices to separate the two. Hence, in driving a car, we maintain postural stability while transporting the body forward. However, it is important to distinguish between functional role and neuromotor control processes to understand task requirements. The action's functional role (not the engagement of the postural or locomotor system) determines the information-processing demands placed on the performer. Hence, driving an automobile requires processing of information in a fashion similar to walking or running. In both tasks, the performer must determine the rate of body transport, scanning distance, and risk-level.

Interaction with Objects

Functionally, many actions involve changing or maintaining the position of objects, contacting other people, or altering the terrain in some fashion. Although such actions can be carried out using any segment of the body, hands are typically employed. In the present discussion, all interactions with objects, other people, or the terrain are referred to simply as "manipulation." However, use of this term is a convenience and does not imply only hand movements. Any body segment may be used to achieve the action's purpose.

Manipulation Absent. When the performer engages in actions that do not involve manipulation, the task simply requires control of body orientation. In such tasks, all body segments, including the upper limbs and hands, are integrated (or "yoked") into the movement pattern used for stability or transport. In Figure 3–3 are several illustrations of how upper limbs and hands become yoked into the postural system providing for body stability. Frequently, stability is enhanced by increasing the number of body segments in contact with a solid surface. Consider such situations as leaning on an elbow while sitting at a desk, holding onto a strap as you stand in a moving bus, or stabilizing sitting in a reclining chair by maintaining contact of all body segments with the support surfaces. During upright stance, simply maintaining contact of one fingertip with a solid support surface has been shown to decrease body sway.[32]

Maintaining body stability is not the only circumstance in which the upper limbs assist in body orientation. Arm swing during locomotion or crawling on all fours are examples of integrating upper limbs and hands into the movement

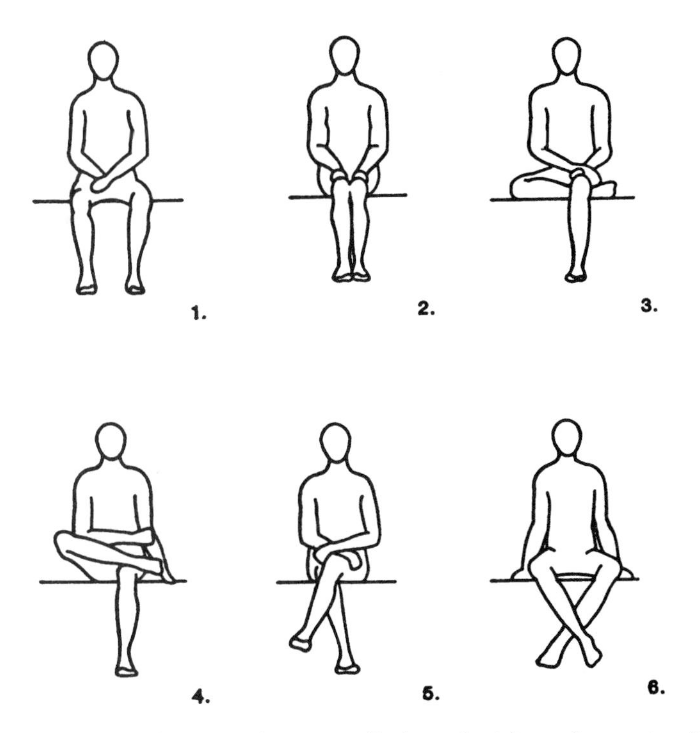

Figure 3–3 Sitting patterns showing the upper limbs yoked into the postural system. *Source:* Reproduced by permission of the American Anthropological Association from American Anthropologist 1955. Not for further reproduction.

pattern used for body transport. When manipulation is not required by the task, the natural engagement of upper limbs/ hands into postural or locomotor patterns seems reasonable. The upright posture of humans, which liberates the hands from stability and transport functions, is a relatively recent development in evolutionary history.

Manipulation Present. Interaction with objects requires freeing some segment of the body from stability or transport functions. As discussed previously, upper limbs and hands are typically freed from postural support or locomotor functions and used to manipulate objects. However, interaction with objects is not limited solely to use of the hands. We can use a foot or the head to impart force to an object (as in the game of soccer). We can hold open a door with an elbow or a knee. Further, individuals without use of both upper limbs can interact with a computer using a stick held in the mouth. Granted, hand movements provide for more precise adjustment of objects than use of any other body segment. However, it is important to recognize that almost any body segment can be recruited for "manipulatory" action. Even trunk motion has been shown to be part of a "reaching" movement when an object's distance exceeds the arm's length.[33]

Using any body segment to successfully interact with objects requires advance and ongoing modulation of neuromotor processes controlling body orientation. Thus, reaching to grasp a cup requires feedforward adjustment of postural support.[34,35] Otherwise, the upper limb motion disturbs body stability. Similarly, running to kick a ball requires feedforward adjustments of locomotion. Later on in this chapter, we discuss the learning processes mediating such feedforward control.

In the present discussion, having to concurrently control body orientation and manipulation raises an important issue. In tasks involving interactions with objects, the performer is doing two things at once. Both body orientation and manipulation necessitate gathering information about regulatory environmental conditions and using that information to organize movement. However, regulatory inputs about the two task components may be located at different sites. For example, suppose you are crossing a busy intersection and, at the same time, are searching in your purse or pocket for a bus token. In this task, you have to scan the street for oncoming vehicles, other pedestrians, and irregularities in the road's surface. At the same time, you are tactually exploring your purse or pocket for an object of a certain size, weight, and texture corresponding to the bus token.

Attentional resources are limited. A performer can simultaneously attend to only a restricted number of events. In the prior example, conditions associated

Table 3–3 Function of the Action: Task Categories and Examples

Body Orientation	*Manipulation*	
	Absent	*Present*
	Body Stability	**Body Stability plus Manipulation**
Stability	• Sit • Stand • Lean on table	• Hold object while standing • Reach for glass while sitting • Writing at a desk
	Body Transport	**Body Transport plus Manipulation**
Transport	• Walk • Run • Crawl	• Carry child while walking • Run to catch a ball • Drive an automobile

Source: Adapted from *Structure of Motor Tasks* (p 12) by AM Gentile et al, Professionale de L'Activite Physique du Quebec, with permission of the author, © 1975.

Table 3–4 Function of the Action: Task Characteristics

Body Orientation	*Manipulation*	
	Absent	*Present*
Stability	• Fixed boundaries • Low information processing	• Fixed boundaries • Modulation of postural system • Doing two things at once • Moderate information processing
Transport	• Expanding boundaries • Moderate information processing	• Expanding boundaries • Modulation of postural system • Doing two things at once • High information processing

with the street and the purse/pocket present the performer with a complex, cluttered array. The performer may have to continuously switch attention between these two sources of information (time-sharing).[36] With high stress on attentional resources, it may be necessary to discontinue one part of the task altogether. The performer may have to stop searching for the bus token until the street is crossed. Alternatively, he or she may have to stand still while searching for the token. Thus, when the functional goal involves body orientation in combination with manipulation, there are dual-task constraints and high attentional demands.

The Structure of Tasks: Functional Role

The second dimension of the taxonomy can now be specified by combining body stability/body transport with the presence/absence of manipulation. Four task-categories are derived by combining these two ways of analyzing the action's functional role. Presented in Table 3–3 are examples of tasks in these four categories: (1) Body Stability; (2) Body Stability plus Manipulation; (3) Body Transport; and (4) Body Transport plus Manipulation.

Task-characteristics within each of the four categories are summarized in Table 3–4. As the task shifts from body stability to transport, note the increased requirements for information-processing. Further, the attentional demands increase when tasks involve manipulation because the performer is doing two things at the same time (dual-task constraints).

The Taxonomy of Tasks

We are now ready to examine the complete task taxonomy based on the environmental context and the action's functional role. Presented in Table 3–5 are the 16 task categories that result from the combinations of these two dimensions (ie, a combination of the 4 x 4 arrays present in Table 3–2 and 3–4).[11] Each of the 16 task-categories poses different demands on the performer in terms of information-processing, allocation of attentional resources, and organization of the movement.

Generally, complexity of the task increases as you proceed diagonally from the upper left to the lower right of Table 3–5 (ie, from Closed/Body Stability tasks to Open/Body Transport plus Manipulation tasks). Let us summarize the requirements placed on the performer for these two extreme task-categories (see also: Table 3–2 and Table 3–4). In Closed/Body Stability tasks, all regulatory

Table 3–5 The Taxonomy of Tasks

Environmental Context	*Body Stability* *No Manipulation*	*Body Stability* *Manipulation*	*Body Transport* *No Manipulation*	*Body Transport* *Manipulation*
Stationary No intertrial variability	Closed Body stability	Closed Body stability plus Manipulation	Closed Body transport	Closed Body transport plus Manipulation
Stationary Intertrial variability	Variable Motionless Body stability	Variable Motionless Body stability plus Manipulation	Variable Motionless Body stability	Variable Motionless Body stability plus Manipulation
Motion No intertrial variability	Consistent Motion Body stability	Consistent Motion Body Stability plus Manipulation	Consistent Motion Body transport	Consistent Motion Body transport plus Manipulation
Motion Intertrial variability	Open Body stability	Open Body stability plus Manipulation	Open Body transport	Open Body transport plus Manipulation

environmental conditions fall within graspable space, are stationary, and do not change from one attempt to the next. The task is self-paced and imposes no predictive demands. With practice, the performer hones in on a stable movement topology and tries to specify similar movement parameters. The movement becomes consistent and habitual, implicating a reproductive mode of organizing neuromotor processes. As the environment is unchanging, continuous monitoring decreases, freeing attentional resources for other activities.

In contrast, Open/Body Transport plus Manipulation tasks require coping with objects or people in motion and with conditions that change over successive attempts. The task is externally paced and poses high predictive demands. Practice diversifies the movement pattern, implicating a generative mode of movement organization. The performer becomes adept at composing and controlling movement patterns to fit environmental variations. Ongoing scanning of the surroundings is required and the performer becomes better attuned to critical environment events. As body transport is involved, boundaries of the environment are constantly expanding. To determine how far ahead to gather usable information from the environment, the performer must continuously evaluate the rate of transport, his or her inherent time-lags, and the risk-level tolerable. Dual-task constraints are imposed on the performer: body transport and manipulation are carried out concurrently. Information regulating body transport and manipulation are likely to be at distinct sites in the environment, increasing the complexity of processing information and taxing attentional resources. In addition, feedforward adjustments of the locomotor or postural system are needed to permit freeing a body segment for manipulation.

To summarize, we have been describing the requirements imposed on the performer by two extreme types of tasks: (1) Closed/Body Stability, and (2) Open/Body Transport plus Manipulation. In a similar manner, requirements associated with the 14 other task-categories (Table 3–5) can be specified by considering prior analyses and by examining the summaries presented in Table 3–3 and Table 3–4.

Practical Applications of the Taxonomy

There are several ways in which this taxonomy might be of value to therapists. First, the taxonomy provides a comprehensive framework that can be used to *evaluate patients*. Tasks can be devised for each of the 16 categories that would be appropriate for a particular patient-population. By observing performance under these different task-constraints, the therapist can determine the degree of complexity that the patient can handle. Further, this type of comprehensive evaluation would be useful in dealing with government or insurance agencies that fund health care. If the patient is only evaluated in Closed tasks or in body orientation tasks that do not involve manipulation, continuing therapy may seem unnecessary. However, these relatively simple tasks are not representative of the daily situations with which the patient must cope. Closed tasks are the least common type of activity in everyday interactions with the physical environment. Therefore, evaluating patients should be based on a broader range of activities, such as the 16 task categories outlined in the taxonomy.

Second, the taxonomy provides guidance in the *selection of therapeutic activities* that are functionally appropriate. Using the taxonomy ensures that therapeutic interventions cover the range of tasks required for effective adaptation to the physical environment. Third, the taxonomy can also be used to *define a patient's performance deficit.* In doing so, the therapist can systematically vary the task structure to determine whether the patient has difficulty with information-processing, attentional mechanisms, movement organization, or motor control systems.

Fourth, therapists can use the taxonomy to *chart progress.* The patient's performance on tasks within different categories provides a profile of competencies in everyday behaviors that can be shared with the patient, caretakers, and the family. Such a profile can provide direction as to which activities are being carried out adequately, need continued improvement, or are beyond the patient's present capabilities. Recently, McCaffrey-Easley[37] used one dimension of the taxonomy, the environmental context, to develop an assessment strategy that yields such a patient profile. She specified several activities within each of the four categories and developed a scaling strategy that can be used to evaluate the capabilities of young children and to chart progress by documenting change.

Last, an important benefit in using the taxonomy is that it provides *insight into skill acquisition processes.* Learning is task-dependent. The processes involved in learning a Closed/ Body Stability task are different from those associated with an Open/Body Transport task. As learning differs, educational or therapeutic practice should also differ. As we discuss subsequently, clinical interventions designed to facilitate skill acquisition should be determined by the task structure.

SKILL ACQUISITION

An Overview

During skill learning, the performer views the environment differently and alters his or her behavior in that context. Let us consider two examples of skilled performance. In the first, the task requires precise knowledge of the environment, high-level predictive processes, and active decision making. However, the movements used are relatively simple for an adult. In the second example, processing of information is somewhat similar to the first but movement organization is highly complex.

Two Examples of Skilled Performance

The Train Driver. The first situation involves the driver of a railway train whose behavior is guided by a goal of having the train arrive safely and smoothly at a station by a specified time. As described by Branton,[38] the skilled driver has a view of the environment characterized by an extended span of foresight. The conditions of the upcoming track and terrain are known to him. His driving behavior is governed by an overall strategy of maintaining the train's velocity within certain limits specified by passenger safety and comfort and by arriving at the station on time. Tactical modifications of this strategy are arranged in advance to cope with local conditions, such as curves, changes in gradients, or upcoming signal posts. This capability to foresee invariant conditions does not imply that the driver fails to attend actively to the track. On the contrary, the skilled driver knows just where to look to pick up cues signaling possible hazards. When confronted with new information, the driver is able to adapt his behavior flexibly and in an unhurried fashion. He has encountered a variety of novel situations in the past and can draw on these experiences to generate a new approach quickly. As noted by Bartlett,[39] the skilled performer appears to have all the time in the world.

This first example highlights some key characteristics of skill learning. These involve memory of invariant environmental features that permit a wider span of foresight, predictive capabilities with reference to conditions that may occur, and effective scanning of the environment to pick up relevant information. The learning process most implicated in this type of skilled performance is referred to as *explicit*. Analyzing and predicting environmental events and selecting appropriate movement-options are consciously available to the skilled performer.[1] These events, decisions, and actions can be described verbally or can be "brought to mind" in the form of visuospatial images. Such learning implicates overt problem solving: hypotheses are generated to guide planning; plans are evaluated on the basis of information-feedback; and active decision making guides subsequent attempts.[40] As a result of explicit learning processes, the performer develops a mapping function (a set of correspondences) between environmental events and his or her morphology yielding movements that successfully achieve the goal.

The Skater. The second example of skilled performance involves a person using "in-line" skates whose goal is to move swiftly and safely over a paved surface. The skilled skater engages in processing information, looking ahead, and making predictions that are, in many ways, similar to the train driver (ie, both tasks involve body transport). Thus, the skilled skater adjusts movement well in advance of having to cope with upcoming events (such as the location of barriers or a change in the pavement's surface). Although these aspects are similar, performance of the two tasks differs in important ways.

The skater's organization and control of movement, involving almost all body segments, is more complex than the train driver's manipulation of throttle and brake. The skater must apply force against the ground that is timed and graded precisely to maximize gliding and minimize energy cost. Maintaining a balanced body position while skating is a challenging task, especially with the rapid rate of forward motion and the minimal base of support provided by the skates. Despite these complexities, the skilled skater's movement is smooth, flowing, and appears almost effortless. Lateral shifts in body position rhythmically offset the propulsive motion of the lower extremities. Changes in direction are set up by smooth transitions, are initiated in ample time to avoid perturbing the movement's flow, and link all body segments in the adjustment process.

This second example reflects a learning process that involves more than matching movement to the environment to transport the body from one place to another. Less skilled performers might manage to move about on skates, might even handle the information-processing demands but, qualitatively, their movements are different. The unskilled skater's movement lacks smoothness and is inefficient. As is discussed in more detail in later sections, refining the movement's organization to increase efficiency occurs slowly over extended practice. The control processes mediating these refinements are not consciously available to the performer. Hence, learning involving these control processes is referred to as *implicit.*[1]

Explicit versus Implicit Learning Processes

Two interdependent processes have been proposed to underlie skill acquisition.[1] These processes operate in parallel, are differentially accessible to conscious awareness, and affect change at different rates as a consequence of practice. An explicit learning process is primarily responsible for the rapid improvement observed during initial practice. The gradual changes seen with extensive practice result from implicit learning.

The combined influence of these two learning processes may underlie the "Power Law of Practice" first identified by Snoddy[41] and later shown by Crossman[42] to hold across a variety of tasks. Basically, the law quantitatively describes the relationship between amount of practice and change in performance: Rapid improvement is seen early in practice with progressively smaller improvements occurring as practice continues over extended periods of time. Previously, several explanations have been advanced to account for the Power Law.[43–45] In the present discussion, a new alternative is proposed. The Power Law is assumed to reflect changes in performance induced at different rates by implicit and explicit learning processes. Explicit learning affects performance changes more rapidly than implicit. Let us now consider these two learning processes in more detail.

Explicit learning leads to increased success in attaining the action-goal. With practice, this learning process gives rise to a set of correspondence or mapping between the performer's morphology and the regulatory environmental conditions. This mapping is reflected in the movement's overall form or topology and in the specification of appropriate movement-parameters that match task-constraints. The performer can consciously guide this mapping function and is aware of changes in the movement's topology and parameters. As a result of explicit learning, a movement pattern stabilizes rapidly during initial practice of a new functional task. This movement is "good enough" to meet task demands and attain the goal; but it is not efficient. As practice continues, explicit learning continues to influence performance and provides for enhanced information-processing, better predictive capabilities, and more precise mapping of movement to the environmental context. Explicit learning enables more consistent success in attaining the action-goal. However, explicit learning processes do not mediate the smooth, flowing movements characteristic of skilled performance.

Implicit learning is responsible for changes in movement organization that lead to efficient performance. Movement organization changes in at least three ways with practice: (1) regulation of intersegmental force dynamics; (2) blending of successive movement components; and (3) coupling of simultaneous components. Customarily, these aspects of movement organization are considered as "motor control" processes. However, as pointed by Willingham,[46] motor control and learning need not be considered as separate events.

Although all three processes are discussed in more detail later in this chapter, let us consider the changes in intersegmental force dynamics that result from practice and, therefore, are due to learning. Efficient control of force dynamics requires the precise balancing of active and passive effects. Muscle contraction provides the active force component. Passive forces include external field effects (gravity, friction, and contact forces) and motion-dependent torques (reactive forces resulting from linkage effects, centripetal, and Coriolis forces). As so elegantly described by Bernstein,[7] the skilled performer learns to exploit these passive forces by timing and grading muscle contraction to exactly "make up the difference" between these passive components and the movement's desired topology. In organizing movement, the skilled performer profitably uses these passive forces that one "gets for free" (ie, without active muscle contraction). Capturing the precise moment for production of muscle force yields efficient motor output and smooths the movement's flow. The performer cannot consciously guide the interplay of active and passive forces. To predict passive effects and optimally regulate muscle contraction, an internal feedforward model is required that uses estimates of the inertial characteristics of body segments and information about the current status of musculoskeletal structures.[47] Practice leads to developing this internal model that mimics the interplay of active and passive forces and provides for the predictive specification of appropriate muscle force.

Practice also produces alterations in other "motor control" functions mediating efficient movement. There is blending of initially distinct movement components within a sequence that involves the anticipatory co-adjustment of successive units. Similarly, practice leads to the temporal coupling of motor components that are carried out simultaneously. We do not voluntarily determine nor are we consciously aware of these changes. Implicit learning involves self-organizing processes that do not reflect the performer's overt problem solving and hypothesis testing. Rather, interactional modes within the CNS appear to be passively reinforced when the motor output minimizes some cost function (eg, energy, time, or the need for information).[48] These organizational changes come about slowly, as a result of extended practice, and are not consciously accessible.

Phases of Skill Learning

Because explicit and implicit processes change at different rates, skill learning gives the appearance of taking place in stages. During initial practice on a novel task, explicit learning processes are most influential. Active problem solving is evident: the learner explores various ways of organizing movement in an attempt to successfully reach the goal. Fitts,[49] a researcher who made important contributions to the study of motor learning, referred to this first stage of learning as "cognitive." In addition to the overt problem solving evident during early learning, adults seem to guide their behavior verbally. Such "talking to one's self" may help to mark key task-features and aid in recall. During this initial phase, the learner discovers a movement topology that matches the environment and successfully attains the goal. However, consistency in attaining the action-goal is not high and performance is not efficient.

As practice continues beyond this initial phase, the performer becomes more adept at tuning the movement to task-constraints by specifying spatial and temporal parameters. The extent to which the movement's topology must be adapted to environmental events is *task-dependent*. For example, Open tasks require specifying a wide range of movement parameters to match the variable regulatory conditions, whereas Closed tasks do not. Learning to parameterize movement is similar to the "associative" processes that Fitts[49] described as occurring during intermediate stages of skill acquisition. Associating movement parameters with specific environmental events leads to increased success in achieving the action-goal.

During later phases of learning, the movement becomes more efficient.[50] Implicit learning processes are most influential at this stage. After extensive practice, the gradual refinement of control processes enables the learner to carry out the movement consistently and smoothly.[50] Fitts[49] characterized this third stage as leading to "automaticity," a description that resembles the automatic and unconscious processes that promote movement efficiency.

Given this brief overview, let us consider learning as it occurs naturally (ie, without instruction by a parent, peer, teacher, or therapist). Subsequently, we will discuss how therapists can facilitate skill learning by appropriate instructional interventions.

Initial Phase of Learning

During initial practice of a novel task, the performer acquires a general concept of the movement that is successful in achieving the goal.[6] Our analysis of early learning involves several component processes. The action-goal establishes an organizational framework for the learner's behavior: it determines which environmental conditions are regulatory and it directs the performer's selective attention. Based on information about the environmental context, the performer plans the initial movement pattern. Execution of this movement produces feedback that enables the performer to decide how to organize subsequent attempts. We now examine each of these components in more detail.

Goal of the Action

Action-goals arise from the motivational state of the performer and the opportunities afforded by the environment. Both factors have a potent influence on how the environment is perceived and how movement is organized. The general class of action-goals that we are considering involves body orientation, manipulation, or both concurrently. How to achieve the action-goal is a problem for the performer: the end is clear, the means is not. Solution of this problem requires identifying movement approaches that produce the desired outcome. Hence, what is acquired during this first stage of learning is a means-end relationship.

Regulatory Environmental Conditions

The goal specifies those environmental conditions that are regulatory. As we discussed previously, when lifting an object, movement must be organized spatially to exactly match the object's dimensions. Similarly, if the action-goal is to step onto an escalator, the escalator's motion and spatial configuration determine the movement's timing and spatial features. Once this goal is established, the person is no longer free to decide when and where to move. To be successful, the performer must mold movement to these regulatory conditions.

The movement pattern that yields the desired outcome is, therefore, under the control of the environment.

Selective Attention

The performer learns to focus on environmental features that are relevant for performance and ignore those that are not. This is not always easy for the learner, especially in complex situations involving many regulatory features embedded in irrelevant arrays. For example, imagine a child in a playground trying to throw a lightweight ball into a basketball hoop. The child detects the size, shape, weight, and texture of the ball and the height, distance, size, and other characteristics of the hoop. All of these conditions are important in organizing the movement. However, over repeated attempts, using a variety of movement patterns, the child is not successful. A crucial aspect of the environment has been overlooked. A brisk wind carries the ball to the right each time the child throws it toward the basket. Until that environmental feature is identified as regulatory, success is not likely.

Most action-goals implicate physically structured entities: other people, objects, surfaces, or the terrain. The performer readily perceives the spatial structure of these entities. However, during early skill acquisition, two types of perceptual learning enhance the identification of relevant information: differentiation and grouping.[51] Differentiation implies that the more familiar you are with a situation, the more detail you tend to see. Initially, the learner may not discern subtleties of the environment that are crucial for organizing an effective movement. For example, beginning skiers are focused on their skis, the terrain, and the upcoming slope. Later, they may begin to notice that the snow's quality influences body stability and rate of body transport. At first, all they see is "snow." With practice, they distinguish qualitatively between different types according to grain, texture, wetness, and slickness. Being able to discern these differences, they can organize movements more effectively. Hence, perceptual differentiation enables the learner to identify and attend to regulatory features and, thus, to better cope with environmental conditions.

The second type of perceptual learning involves grouping of information into larger units, which is often referred to as "chunking." This type of learning is particularly necessary when flow of information is extended in time, as is the case when objects or people are in motion. For example, in learning to drive an automobile, it is essential to detect flow patterns in the motion and paths of pedestrians and other vehicles. Rather than seeing each car's motion as an independent event, practice enables the driver to pick up the relationship covering several vehicles. Motion of several cars becomes grouped in time into a single event. Perceptual grouping of environmental events reduces attentional demands and frees resources to be applied to other aspects of the task.

Planning the Movement

When first attempting a novel task, the performer plans a movement using whatever resources are available from prior learning or through intrinsic organization. The learner must solve several problems as he or she tries to assemble an appropriate movement pattern. These problems concern different aspects of planning and are associated with (a) the movement's topology, (b) parameter specification, (c) simplification, and (d) motor control.

Topology. Topology is the general, configurational aspect of the movement that remains stable although the pattern may be scaled up or down in amplitude, rate, or force. As discussed earlier in this chapter, the movement's topology results from the interaction between two structured entities: the environment and the performer. Movement topology emerges from a mapping between regulatory environmental features and the performer's morphology.

The notion of movement topology was first suggested by Bernstein[7] and has commonalities with the concept of affordances introduced by Gibson.[31] Complementarity of the person and the environment underlies Gibson's definition of affordances. Complementarity implies that physical properties of the environment are perceived with reference to the individual's body structure, posture, and movement. For example, a surface is seen as affording support if it is horizontal, flat, rigid, and extended sufficiently to accommodate the person's size and weight. As described by Gibson, a surface having these four properties is "walk-on-able" or "run-over-able." Solid surfaces raised above the ground to knee height are seen as "sit-on-able." Small detached objects are graspable with one hand; larger ones are liftable with two. Gibson was concerned with the perceptual processes through which an individual detects the affordances provided by the environment. The concept advanced here is that such perception occurs concurrently with the shaping of the movement's topology.

The action-goal determines which movement afforded by the environment takes precedence at any given time. If the performer is distracted from the task, environmental affordances may capture the behavior. Recording such slips of action, Reason[52] reported the following experience: "I meant to get my car out, but as I passed through the back porch on the way to the garage I stopped to put on my Wellington boots and gardening jacket as if to work in the garden." In this example, attentional drift allowed momentary affordances of the environment to take control of the behavior's direction. Maintaining focus on the action-goal frees the individual from the dictates of the environment and directs appropriate selection from the various movement options that the surroundings afford.

The movement's topology is available to conscious awareness. The performer may, however, selectively attend only to certain aspects of the movement that serve as "markers" or "turning points." Markers are the intermediate and end-point configurations of body segments taken with reference to objects and surfaces. Turning points specify transitions within a temporal sequence. For example, in stepping over an obstacle, attention is typically given to the path of the working-point, the foot, and especially the key markers of toe-clearance above the obstacle and of foot-placement at the end of the step. These markers capture the performer's attention as they relate most directly to the action-goal of safely clearing the obstacle.

Thus, certain body parts have "star billing," while other body segments serve as "supporting players" (to use Brooks' colorful metaphor[53]). In picking up a cup, the hand is the focus of attention (ie, the star) while trunk and lower limbs serve a supporting role of maintaining body stability. In contrast, trunk and lower limbs take star billing when standing up. Now, the upper limbs and hands are delegated to a supporting role, perhaps by exerting force against the chair's arms.

Parameter Specification. Specifying the force, time, or amplitude of movement is a second aspect of motor planning. Often, this process is referred to as setting the parameters within a motor "program." Varying a movement's temporal or spatial parameters does not alter its topology. Using handwriting as an example, Viviani and Terzuolo[54] have shown that the configurational aspects of signing one's signature remain relatively the same although the writing is varied in size and speed. Further, parameter specification appears to happen at the same time but independently of the configurational aspects of the movement. Research has shown that choice of responding arm and movement direction (configurational aspects) and movement amplitude (parameter) are resolved in parallel.[55] These findings suggest that a movement is not planned serially, ie, topology is not organized before specifying extent. Rather, movement topology and parameter specification seem to be determined concurrently.

Simplification. The need to simplify movement organization was first noted by Bernstein[7] and is referred to as the degrees of freedom problem. As so aptly described by Turvey and colleagues,[56] the problem in planning movements is that "there are too many individual pieces of the body to be regulated separately." The complexity of assembling a movement from individual "body pieces" such as joints, muscles, or motor units is enormous. Consider the difficulty in organizing an arm movement if the motor plan involves commands to individual joints. Movement at the shoulder, with the arm fully extended, can occur on three axes: horizontal, vertical, and longitudinal. To control position, values on these three axes have to be specified; hence, there are three degrees of freedom. The radial ulnar joint permits rotation around its length, thus adding one more degree of freedom. One degree of freedom is contributed by the elbow and two by the wrist (horizontal and vertical movements). Therefore, motor plans consisting of commands to individual joints would require specifying seven values and the changes in these values over time.

The situation is even more complex if a motor plan is envisioned as commands to individual muscles. Ten muscles act at the shoulder joint, four at the radial ulnar, six at the elbow, and six at the wrist. Movements organized by commands to muscles would have 27 degrees of freedom, ie, specification of 27 values would be required. Obviously, the number is larger if commands are directed to individual motor units. This analysis considers only the complexity in controlling arm motion. However, the arm does not move in isolation but involves concomitant adjustments of those joints, muscles, and motor units controlling motion across all body segments. Viewed from this perspective, the degrees of freedom to be regulated are too numerous for motor plans to be based on commands to separate entities.

To simplify planning during early learning, degrees of freedom are limited. This can be accomplished in two ways.[7] First, brute restrictions can be imposed on joint motion of all but the most essential body segments. "Freezing" degrees of freedom by limiting joint motion is achievable by co-contraction of antagonist muscles that increases joint stiffness. However, this strategy may yield movements that are stiff and rigid. Further, organizing movement by markedly restricting joint-motion may limit success in attaining the goal. Research suggests that freezing degrees of freedom is used only during the initial attempts at a novel task. For example, using a slalom-like ski task, Vereijken and colleagues[57] found that learners restrict joint motion primarily during the first few trials, after which variability increases with practice. Therapists should be aware that co-contraction is a typical response in unimpaired individuals during initial acquisition of a novel task. Hence, patients should be expected to display a similar strategy during early learning.

The second way in which planning can be simplified is by organizing neural processes that link motion at two or more joints or link muscle groups spanning several joints.[7] Such coordinative units do not yield one fixed pattern of movement. Rather, they establish a mode of interaction among many neural subsystems whose concerted action leads to the specification of an appropriate movement. Russian investigators[58] have introduced the term "synergies" to describe the coordinative units underlying movements with similar kinematic features. These investigators suggested that basic synergies are inherent to the organism and provide the building blocks for the acquisition of new functional linkages. Other synergies are acquired and represent abstractions from highly overlearned patterns of movement.[58] Synergies specify relationships within a set of joint motions or across a group of muscles. Even though the number of joints or muscles may be large, simplification is achieved because the set has only one degree of freedom. Processing of sensory feedback resulting from the movement is also simplified as it is organized with reference to the functional relationship defined over joints or muscles.

For example, locomotor synergies simplify the organization of walking. They provide for the repetitive, rhythmical activation of limb muscles. Consider movement of the lower limb during the step cycle. Ordering of ankle, knee, and hip motion is stable regardless of the speed of walking. Central pattern generators (CPGs) have been identified as the mechanism underlying this orderly sequence of joint motion.[59] Oscillatory networks, apparently involving interneuronal networks at the spinal level, are implicated as the neural substrate for locomotor CPGs.[60] Feedback from the moving limbs and inputs from other subsystems modulate the output of these locomotor CPGs, giving rise to the detailed shaping of the step cycle. Locomotor CPGs simplify the planning process. Thus, the basic stepping sequence does not have to be composed anew for each instance of walking.

Although coordinative units simplify the planning process, skilled performance necessitates the parallel operation of other neural subsystems. For example, skilled walking requires more than simply controlling the stepping sequence. Balance must be preserved during locomotion by regulating force dynamics across all body segments. The step must be parameterized based on information about the ground's surface so that precise foot-placement is achieved. Furthermore, visually steering a path through a cluttered environment is required. Beyond the CPG, control of these locomotor components is mediated by several neural subsystems located at different sites within the CNS. Thus, skilled performance depends on ongoing interaction among all of these distributed neuromotor processes.

Control Processes. During initial learning, the performer's movement may "get into the ballpark" (to use Greene's[61] metaphor) and successfully attain the goal. However, the movement reflects only a crude organization of the underlying control processes. As discussed earlier in this chapter, movement emerges from the dynamic interplay between active and passive forces. Muscle contraction provides the active force component; passive forces arise from the external field and from motion-dependent effects. For example, linkage of skeletal structures produces mechanical interaction forces at joints distant from the sites of active motion. Within multilinked kinematic chains, acceleration of motion at one joint passively produces torques at remote joints. These intersegmental effects, along with other motion-dependent forces, disrupt the movement's flow unless they are accounted for in advance.

During initial practice, the performer has not developed an internal control model that predictively accounts for the interaction between these active and passive forces. Thus, passive forces are not anticipated and give rise to jerky and jagged movements. Performance appears uncoordinated and poorly controlled. As was the case in "freezing degrees of freedom," the novice performer may use co-contraction of opposing muscle groups to increase joint stiffness and reduce these perturbing effects. Although not an efficient strategy, co-contraction does prevent major distortions in the movement's topology.

The failure to precisely control active and passive forces leads to inconsistent movement patterns. Variability also results from the novice's inability to smoothly mesh movement components performed successively or simultaneously. However, such variability should not be viewed as error. As pointed out by Kelso and Ding,[62] it is the natural consequence of a dynamic and self-organizing system. During initial learning, movement variability is the outward manifestation of exploration and search among organizational options: it is essential for the refinement of control processes. Thus, variability is necessary for implicit learning processes that will eventually, over prolonged practice, enable efficient movement patterns to emerge.

Performance of the Movement

The processes affecting movement before and during performance are summarized diagrammatically in Figure 3–4. The action-goal defines the regulatory environmental conditions and guides selective attention to these events. Afferent flows provide information about the environmental context and about the

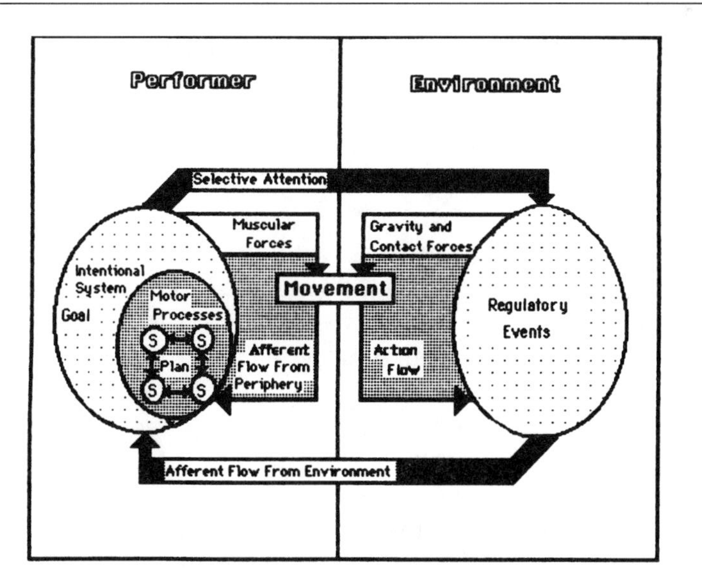

Figure 3–4 Processes operation before and during the emergence of a movement (see text for discussion).

present status of the body. This information modulates activity within and across neuromotor subsystems ("S" in Figure 3–4), leading to the consensual specification of a motor plan. The movement that emerges reflects the mapping between two structured entities (the performer and the environment) and reflects the interplay between active and passive forces (shown in Figure 3–4 are only two of these passive effects, namely "gravity and contact forces"). Movement changes the performer's relationship to the environment (ie, movement produces an action-outcome). Movement also changes the afferent flows that now provide information feedback about the action-outcome and the movement's execution.

Feedback

Information that occurs during and after the performance, as a natural consequence of behavior, is referred to as intrinsic feedback. Two types can be distinguished: (1) information feedback about the outcome; and (2) information feedback about the movement. Outcome feedback indicates whether the goal is achieved and is often termed knowledge of results. For example, in throwing a dart at a target, visual feedback enables the performer to determine whether the dart hits the target. In picking up a cup, cutaneous and visual feedback provide information about the outcome. Although any sensory modality can provide input indicating the action-outcome, we tend to rely on visual input in our everyday interactions with the environment. The second type of feedback signals how the movement was performed. Such feedback is sometimes referred to as knowledge of performance. Several sensory modalities contribute to this afferent flow, especially proprioception (coming from vestibular, joint, and muscle stretch receptors).

Information feedback is available for only a brief period of time (of the order of seconds). It must be actively categorized and coded to be retained longer. We do not retain specific sensory consequences. Instead, inputs about the outcome are organized with reference to the goal. A minute or so after the action, we recall "hitting the target with the dart" or "missing the target by hitting too far to the left." Information about movement's execution is also organized and coded. Thus, sensory input from the movement is partitioned into relevant sets for use by neural subsystems responsible for specific aspects of motor planning.

If movement is executed slowly, information feedback may be used to correct or modify the ongoing performance. Often, however, there is insufficient time to make such adjustments. Earlier, we discussed the time-lags in processing information. When these time-lags do not permit corrections during the movement's execution, feedback is used to update the motor plan for the next attempt. However, the motor plan is not changed on the basis of movement feedback alone. In making decisions about the next attempt, the learner considers information about both the outcome and the movement.

Decision Making

Explicit processes guide movement planning for subsequent attempts. The learner consciously accesses an internal representation of the movement's topology and parameters, as well as information feedback about the movement's performance. In addition, the action-goal and feedback about the outcome are consciously available. A means-end analysis guides decisions as to how the next movement is to be organized. In this approach to problem solving, a desired state (the action-goal or the motor plan) is compared with relevant feedback. Detecting a difference between the desired end-state and feedback leads to some modification in the movement. In a sense, this process represents hypothesis-testing: "If movement is organized this way, then the end-state will be achieved." This hypothesis is "tested" by comparing the intended state with feedback.

In applying this problem-solving strategy after an initial attempt at the task, the most immediate concern of a learner is: "Did I accomplish the goal?" To answer this question, the learner compares the action-goal with information feedback about the outcome. This comparison requires that both sources of information are held in short-term, working memory. Barring neurological damage or severe distraction, the action-goal is usually remembered. However, slips in memory do occur, even in unimpaired people. You may have had the experience of opening a drawer and not remembering the purpose for doing so. You shuffle items in the drawer hoping that seeing one of them will reinstate the action's original purpose. Clearly, you have feedback about the environment: the drawer is open and its items are visible. However, information feedback about the outcome is not sufficient to determine whether you were successful in the task. The goal guiding your action must be retained.

The performer's other concern is: "Did I move as planned?" To answer this question, the learner compares the consciously available components of the motor plan (topology and parameters) with feedback about these movement aspects. Again, both types of information must be available in working memory for this comparison to take place. Feedback alone is inadequate to determine whether the movement was executed as planned.

On the basis of these two comparisons, the learner decides what to do next. For simplicity, the results of these comparisons are dichotomized into simple yes/no categories (see Exhibit 3–1). Let us consider the case in which a "yes" is obtained for both comparisons. The performer achieves the goal by moving as planned. Success in achieving both the action-goal and the intended movement increases the likelihood of repeating the same approach. Thus, the learner typically maintains the same motor plan for the next attempt. In effect, the

Exhibit 3–1 Decision Processes

		Did I Accomplish the Goal?	
		Yes	No
Did I Move As Planned?	Yes	A Successful Solution	Something's Wrong
	No	Surprise!	Everything's Wrong

Source: Reprinted with permission from A.M. Gentile, A Working Model of Skill Acquisition with Applications to Teaching, *Quest*, Vol. 17, pp. 3–23, © 1972, Human Kinetics.

motor problem has been solved. The movement's topology was shaped to the environment and movement parameters matched task-constraints. Therefore, the desired outcome was produced. The learner might say, "I have the idea of the movement," indicating that an effective means-end relationship has been identified. The learner has developed a general concept of an appropriate movement.

Often during early learning, the movement is performed as planned but fails to attain the goal. The learner decides that "something's wrong" in the way movement was organized (see Exhibit 3–1). The movement's topology or parameters did not match the task-constraints. This outcome requires that the environment be reassessed and the motor plan revised. When comparisons yield "yes" for the movement and "no" for the outcome, the learner usually modifies some aspect of the movement on the next attempt.

Sometimes, the action-goal is reached even though the movement is not performed as planned (ie, the emergent movement's topology and/or parameters differ from those specified). During initial learning, control of intersegmental force dynamics is crude and passive forces frequently perturb the intended movement. Occasionally, just by chance, these perturbations produce a movement pattern that matches the environment and successfully achieves the action-goal. One would think that learners would try to reproduce this aberrant but successful movement. However, in a study of decision making during skill learning, Skinner[63] found that performers often do not attempt to reproduce this movement. Learners seem unable to inductively construct a motor plan based solely on feedback about the aberrant but successful movement pattern. Apparently, such feedback has relevance only when compared with the original plan. In any case, the most common strategy on the next trial is to try to produce the movement that was planned originally.

After an initial attempt at the task, the outcome may be that "everything's wrong" (see Exhibit 3–1). The learner fails to produce the intended movement and the action-goal is not attained. Following this outcome, several options are available to the performer. The learner can try the intended movement again or revise the motor plan. With repeated failure, the learner might change the goal or even quit. How performers handle this situation probably reflects their motivation level and their prior history of success and failure in similar tasks. The unimpaired adolescent boys in Skinner's[63] study chose to retry the original plan. We have no information as to how impaired individuals respond in this situation. It is not unreasonable to think that, without the encouragement of a therapist, some patients would quit or change the goal.

During initial practice, different ways of performing the task are tried until the learner has some success in achieving the goal and in executing the movement as planned. Although the learner now has a general concept of an effective approach, he or she is not skilled. The action-goal is not achieved consistently and the movement lacks efficiency. Skill acquisition occurs during later phases of learning as a consequence of continued practice.

Later Phases of Learning

Successful Action

Increasing success in achieving the action-goal depends on meeting challenges that are task-specific. The taxonomy, presented earlier in this chapter, outlined 16 types of tasks that impose different requirements on the learner. With practice, the learner becomes more proficient in coping with these task-constraints. Practice changes how information is processed and how movement is organized. Input and output streams are organized into higher-level units (groupings or coordinative structures) that provide expedient solutions for perceptual analysis and motor control. These changes, leading to more successful action, are now summarized with reference to different types of tasks.

Information-Processing. In Closed or Consistent Motion tasks, information-processing demands decrease with practice as conditions do not change. The environment becomes highly predictable. Thus, the need to continuously pick up information from the surroundings is reduced. In contrast, Open and

Variable Motionless tasks require ongoing monitoring of the environment. Picking up information about changing conditions is essential in these tasks but time constraints differ. In Variable Motionless tasks, the performer can take all the time necessary to analyze the situation. However, Open tasks require coping "on the spot" with motion in the external environment. With practice of Open tasks, the performer learns to identify advance cues signaling upcoming events. In addition, probability functions are developed enabling the performer to evaluate how likely it is that particular events will occur. Using advance cues and probability functions leads to improved predictions about the variable motion characteristics of environmental events and, therefore, results in increased success in attaining the action-goal.

In tasks requiring Body Stability, practice results in performance becoming better attuned to environmental circumstances, especially to the characteristics of surfaces providing support. With practice, we learn to maintain stability under diverse conditions. For example, we become skilled as passengers in moving vehicles: we accommodate sitting to the lurching motion of a bus and, if we are standing, we anticipate and adjust to upcoming changes in the bus's motion. Similarly, in Body Transport tasks, information-processing improves with practice. We learn to precisely adjust scanning distance to the rate of transport and to the risk-level tolerable. In addition, we learn to vary transport rate in accord with access to information about the upcoming environment and based on the time needed to modify the movement.

Information-processing also improves with extended practice in tasks that combine Manipulation with Body Stability or Body Transport. These tasks require doing two things at once and, therefore, tax attentional resources. With practice, the performer learns how to allocate attentional resources and how to pick up information needed for each task component. In addition, time-sharing strategies improve so that the performer knows when and how often to shift attention between the task's body orientation and manipulative components. To meet the dual-task constraints, therefore, learning leads to more efficient strategies for gathering information and allocating attentional resources.

Movement Organization. How the movement changes with practice depends on the type of task being learned. In Closed or Consistent Motion tasks, practice results in movement fixation. Early on, topology stabilizes. Subsequently, the performer identifies the movement parameters that best match the unchanging environment. With practice, the movement shapes more precisely to task-constraints and is increasingly successful in reaching the goal.

In relatively simple Closed tasks, practice leads to movement consistency. For example, Young and Marteniuk's study[64] used a kicking task to examine changes in movement organization with practice (256 trials during one practice session). They minimized the subject's need to maintain body stability while standing by having them lean against a supporting post. In addition, the subject was strapped to the post (strap placed around the mid-trunk). To further secure body position, the subject was also instructed to reach behind and hold onto the post. The kick was executed from a starting position (foot on the floor) to a raised target that required circumventing a barrier placed below the target. The experimenters imposed a timing constraint: the movement had to be completed in 400 msec. Under these restricted task conditions, practice led to a stereotyped movement pattern.

Young and Marteniuk's task markedly reduced the complexities associated with control of intersegmental dynamics and coupling of simultaneous components (body stability and kicking).[64] Without the external postural supports provided in the experimental task, reactive forces would have radiated back from the kicking leg to distant joints in both lower limbs and trunk, disrupting body stability and disturbing the foot's trajectory. Thus, the stereotyped pattern they observed after minimal practice was a function of the task's limited coordinative demands. During early practice of more complex Closed tasks, typical of everyday functional activities, topology may be consistent but performance lacks consistency. As we discuss in detail subsequently, feedforward control, necessary for skilled performance, is based on coordinative processes that are not available to the beginner. The novice's movement pattern is highly variable. Smooth, flowing, and consistent movement in Closed tasks, characteristic of efficient performance, is a function of more extended practice.

In Open and Variable Motionless tasks, in which the environmental context changes from one attempt to the next, movement diversifies with practice. The performer learns to specify movement parameters that precisely match the changing conditions. Topology is maintained but movement patterns are differentiated so that their spatial and temporal parameters fit the variable regulatory events. Along with this ability to differentiate movement patterns, there is increasing success in achieving the action-goal.

Adapting movement patterns to variable environmental constraints can be consciously guided by the performer and, therefore, implicates an explicit learning process. However, attentional demands are reduced as learning progresses. Because only minimal attentional resources are needed, the performer may seem "unaware" of specifying the movement's parameters. For example, adults have had extensive practice adapting their grasp to objects of different sizes. Thus, the recent research findings of Gentilucci and colleagues[65] are not surprising. They reported that adults modify their grasping pattern to match variations of an object's dimension without "noticing" the change. However, failure to fully attend to the details of performance does not imply an unconscious process. When other subjects in their experiment were informed that the object's size would vary, they modified the grasp pattern more precisely and rapidly. Clearly, the processes associated with detecting change and adapting the movement were consciously available to these subjects (ie, they were not implicit). When necessary, the performer can attend more closely to the task and consciously monitor the specification of movement parameters. Therefore, this process does not involve implicit learning.

Similarly, implicit processes should not be assumed to mediate improved performance on experimental tasks that involve a recurring pattern (like tracking or serial reaction time).[66–68] Adults have had extensive practice in detecting patterns in environmental events and specifying movement parameters. Learning of diverse Open and Variable Motionless tasks reduces (but does not eliminate) the attentional demands in coping with environments that change in some predictable fashion. Thus, the failure of experimental subjects to "notice" such patterns in tracking or sequential reaction time tasks may simply indicate that minimal attention is required for detection. If the performer is instructed that a pattern exists, they allocate more attentional resources and are immediately "aware" of the pattern. Thus, information about external patterns can be accessed consciously. In experimental tasks involving tracking or sequential reaction time, subjects often report awareness of a recurrent pattern without being instructed that a pattern is present. Thus, learning to specify movement parameters to fit task constraints in such tasks is mediated by explicit not implicit processes.

In tasks involving Body Stability or Body Transport plus Manipulation, the important consequence of later learning is that separate movement components appear to be reorganized and brought under the control of a higher-level, coordinative unit. With practice, postural support or locomotion is smoothly integrated with motion of the upper limbs and hands. Body orientation and manipulative movements flow together precisely in time. Meshing of these components into one overall pattern simplifies motor control. This reorganization process is probably not mediated by explicit processes. Rather, it is more likely that implicit processes associated with movement efficiency are implicated in these changes.

Efficient Movement

Success in achieving an action-goal occurs well in advance of attaining movement efficiency. Implicit processes mediating movement efficiency change at a slower rate than explicit processes leading to successful goal-attainment. Only with prolonged practice does the performer develop movement patterns that are smooth and efficient. Changes in the movement's organization involve refinement of minimally three motor control processes: (1) regulating intersegmental force dynamics; (2) blending successive movement components; and (3) coupling simultaneous components. Refinement of these control processes takes place outside of conscious awareness. Implicit learning processes producing these refinements seem to occur "automatically" through self-organization of internal interactions. It has been suggested that the organizational modes leading to efficient movement are passively reinforced when input-output relationships minimize some cost function (eg, energy).[48]

Intersegmental Force Dynamics. As discussed previously, efficient control of intersegmental dynamics involves the precise interplay of (a) active forces produced by muscle contraction and (b) passive forces associated with the external field and with motion-dependent effects. With practice, the performer develops an internal model that can simulate (mimic) and, therefore, predict the interaction of active and passive forces.[47]

Predictive control is necessary so that passive effects are accounted for in advance and do not disturb the movement's flow. Reactive forces produced by contact with external objects or surfaces must be anticipated to minimize disruption of the intended movement and carry out an action successfully. For example, when catching a falling ball, research has demonstrated that adults precisely set hand impedance before contact with the ball.[69] They do so by predictively timing and grading muscle activation according to the size of the ball and its falling distance. Thus, adults anticipate the passive effects induced by ball-contact and organize active muscle force to precisely dampen the shock of impact.

Advance control of force dynamics is also necessary when landing from a jump. Anticipatory activation of muscle force in the lower extremities is required to efficiently absorb impact and restore body stability. When jumping from various heights or landing on surfaces that differ in compliance, skilled adults adjust the timing and amplitude of muscle activation.[70] Such predictive modifications of landing patterns imply a feedforward control process that takes into account gravitational, reactive, and intersegmental effects and modulates active muscle force before contact with the ground.

Consider another example of feedforward control: the coordination of postural support and voluntary arm movements. Rapid raising of the arm upward is typically preceded by short-latency activation of postural muscles in the lower limbs and trunk.[71-73] Activation of these postural muscles opposes the mechanical linkage effects produced as a consequence of the rapid arm movement. As discussed earlier, motion at one joint sets up interactional forces that ripple through the multilinked kinematic chain, passively producing torque at remote joints. These intersegmental effects resulting from the rapid arm movement would perturb body stability unless postural adjustments are resolved in advance.

In addition to counteracting perturbations, development of feedforward models allows the performer to exploit passive forces. As pointed out by Bernstein,[7] "...the secret of co-ordination lies not only in not wasting superfluous force in extinguishing reactive phenomena but, on the contrary, in employing the latter in such a way as to employ active muscle forces only in the capacity of complementary forces" (p 109). To move efficiently, the performer learns to capture the precise moment at which active muscle contraction will sum with passive force components (Bernstein's reactive phenomena). Thus, muscle activation is graded precisely to exactly "make up the difference" between passive components and the movement's intended topology. For example, during walking and running, adults use gravitational and motion-dependent forces during the first-half of the lower limb's swing phase, thus reducing the need for active muscle forces.[74,75] Similarly, when adults step over obstacles during locomotion, they exploit passive interactional effects between lower-limb segments to achieve hip and ankle flexion during the swing phase and, therefore, conserve energy.[76]

During initial learning, organization of force dynamics is based on whatever resources are available to the performer. In infants, these resources are limited to endogenously generated patterns subserving spontaneous movement.[77] Later in development, these resources involve previously learned modes of organization (ie, acquired synergies[58]). In either case, when first performing a novel task, organization of intersegmental force dynamics is assembled expediently and results in a crude and inefficient movement pattern.

Success in attaining the action-goal develops much sooner than efficient control of intersegmental force dynamics. The acquisition of reaching during infancy and early childhood illustrates this point. In the earliest attempts to direct the hand to an object, the infant appears to "carve out" a roughly organized movement from endogenously available patterns.[78] At about 4 months of age, the infant's functional reaches are characterized by irregular hand paths, made up of sequential acceleration/deceleration segments (movement units).[79,80] Two months after onset of reaching, the infant is increasingly successful in acquiring the object. However, it takes more than 2 years of practice before the infant gradually gains control over the movement. At this point, hand paths begin to straighten; movement time and number of movement units decrease.[81] However, considerably more practice is necessary before the infant can exploit passive forces in organizing their reaching movements. Konczak and Dichgans[81] observed that infants between 24 and 36 months of age are just beginning to organize elbow extension during the reach by exploiting gravitational and intersegmental linkage effects (the pattern evident in adults). Even at this time, however, the infant's movement lacks the consistency and smoothness of the adult's pattern. Although the child is successful in transporting the hand to the object, the feedforward model is not fully refined. Because feedforward control requires estimates of the body's inertial characteristics, as well as ongoing information specifying body status, the young child's internal model must be constantly updated as growth alters muscle and skeletal structures.

Refining and updating internal models of force dynamics happens gradually throughout childhood as a result of practice and, therefore, learning. Young children often display a movement topology that is similar to the adult pattern and they are increasingly successful in attaining the action-goal. However, their movements are inefficient and inconsistent. For example, the gait pattern of young children approaches an adult-like form at about 6 years of age. During walking on a treadmill, the average profiles for angular displacement at ankle, knee, and hip of children are similar to those displayed by adults.[82] However, children lack consistency from one attempt to the next (coefficients of variation are high).[82] Similarly, when adaptation of gait is required (as in stepping over obstacles), 6-year-old children are typically successful in clearing the obstacle.[83] In addition, both children and adults display similar mappings of obstacle and foot: they scale amplitude of toe-clearance (a key marker within the movement's topology) as obstacle-height increases.[83] In controlling the foot's trajectory, however, these young children display much higher variability than the adult.[83] Thus, 6-year-olds appear unable to precisely modulate active and passive force components to produce consistent movement. For these young children, the movement's topology is stabilized; refinement of intersegmental force dynamics is not. Similar findings of high variability but adult-like topology have been found when young children ascend a step,[84] execute a standing long or vertical jump,[85] and transition from sit-to-stand.[86]

When adults acquire a novel task, movement variability is also high. For example, McDonald and colleagues[87] examined adults learning to throw a dart at a target using their dominant and nondominant limbs (250 trials per day for 10–14 days). The movement's topology (as represented by hand trajectory) stabilized early during practice for both limbs. Subjects linked motion at the shoulder, elbow, and wrist in an attempt to simplify control. However, high variability was evident in joint motion of the nondominant limb throughout practice. The dominant limb displayed similar variability initially that, however, decreased as practice continued. Moore and Marteniuk[88] produced similar findings in an upper-limb aiming task. Initially, adult subjects used co-contraction to simplify the control problem but still displayed high variability in both the myoelectric pattern of muscle activity and kinematic measures. With practice, variability in these movement measures decreased.

Schneider and colleagues[89] examined whether adults learned to use passive force components (ie, gravitational and motion-dependent forces) during practice of a novel task. Using their nondominant limb, seated subjects had to move a handle upward and downward to contact targets as rapidly as possible. A curvilinear movement was necessary as subjects had to clear a horizontal barrier (located midway in the motion). During and after practice, the interplay of active and passive force components was analyzed. The investigators found that gravitational and motion-dependent forces were exploited, especially at the reversal point between the upward and downward motions. These changes in intersegmental dynamics were most pronounced at the shoulder joint. However, with more practice, passive effects were used to advantage at all joints, especially midway through the motion as the barrier was cleared. In a subsequent study using the same task, these changes were shown to minimize the movement's energy cost.[90] These findings demonstrate that adults learn to exploit intersegmental dynamics during practice of novel tasks, which leads to the development of efficient movement.

Let us summarize the major points made in this section. First, variability is characteristic of early learning, as shown by the movements of infants and young children and by adults performing novel tasks. Second, movement variability decreases during learning. With extended practice, the skilled performer develops an internal model of intersegmental dynamics that provides for anticipatory control minimizing the perturbing effects of reactive forces and for exploitation of passive components leading to efficient movement. These changes are not consciously available to the performer. We cannot consciously control the interplay of active and passive force components. Thus, refinement of intersegmental dynamics, which yields efficient movement, is a result of implicit learning processes. However, movement efficiency involves other motor control processes, specifically, the coordination of simultaneous and successive movement components. These control processes are discussed in the next sections.

Blending Successive Movement Components. Early in learning, components within a sequential movement pattern are discernible as separate units. Temporal blending occurs later with more practice. Upcoming movements begin to take shape early in the sequence and components lose their isolated identity. For example, in skilled speech, a phoneme is not produced by one fixed position of articulators. The position of articulators is influenced by phonemes that precede and follow in the sequence. Thus, as described by Ivry,[91] "the point of closure in the vocal tract is much farther forward when pronouncing the /k/ sound in the word 'key' than in the word 'caulk' due to the articulatory gestures required for the following vowel" (p 316). Such preshaping of the vocal apparatus as a function of upcoming and prior phonemes is known as co-articulation.

Anticipatory adjustments are also made in functional movements. Blending of sequential movement results in a more unitary organization and yields the integrated flow characteristic of efficient performance. For example, the skilled typist preshapes the hand for an upcoming letter while the other hand is striking the present letter.[92] As another example, consider the task developed by Johnels and colleagues[93] to assess the temporal integration of movement components in patients with Parkinson's disease. The task requires the individual to bend forward while standing, pick up a small box, walk a few steps forward, and place the box on an elevated shelf. Unimpaired adults overlap these components in time. They complete lifting the box while stepping forward and they begin to raise the box upward toward the end of walking in anticipation of shelf-placement. Overlapping of successive movement units yields a smooth flowing pattern. In contrast, the performance of patients with Parkinson's disease (not on medication) is characterized by far less overlap between successive components.[93]

The blending of components displayed by the skilled typist or by unimpaired adults in Johnels' task implicate an internal representation of the entire sequence. This internal model provides for anticipatory priming and activation of forthcoming components in sequential movements. Some forms of temporal blending (such as co-articulation in speech or hand-preshaping in typing) involve implicit learning and reflect the refinement of underlying control processes.

However, other forms of anticipatory organization appear to be consciously available. Rosenbaum and colleagues[94] have experimentally demonstrated a forward planning strategy in which the way a movement is shaped at the beginning of a sequence sets up optimal conditions for later on. In addition to their experimental task, they described an everyday example of a skilled waiter who intends to pour water into a glass that is initially in an inverted position (with its opening facing down). The waiter picks up the glass by configuring the grip in an unusual manner. In grasping the glass, the hand is rotated with the thumb positioned downward. This unusual grip provides the optimal and least awkward position for pouring when the glass is lifted and rotated so its opening now faces upward. Such forward planning appears to reflect changes in the movement's topology and to be consciously available.

Coupling Simultaneous Movement Components. Another change associated with the acquisition of movement efficiency is the temporal coordination of simultaneous components. For example, in prehension movements, coordination of two components is required: (1) reaching—which involves arm motion to transport the hand to the object; and (2) grasping—which involves finger extension (preshaped to fit the object's features) followed by finger flexion to acquire the object (grip-closure). Grasping in adults is temporally coordinated with reaching so that maximum grip aperture (distance between the thumb and index finger) occurs at about 65%–70% of the reach's movement time.[95,96] In addition, the deceleration phase of the reach is lengthened when the grasp requires high precision.[97] Reach and grasp components are purportedly controlled by two distinct neural subsystems[98] that are loosely linked within a temporal macrostructure, functionally assembled to meet the demands of the task.[99,100]

Recently, Kuhtz-Buschbeck and colleagues[101] have examined how the temporal coupling of reach and grasp develops during childhood. In Figure 3–5,

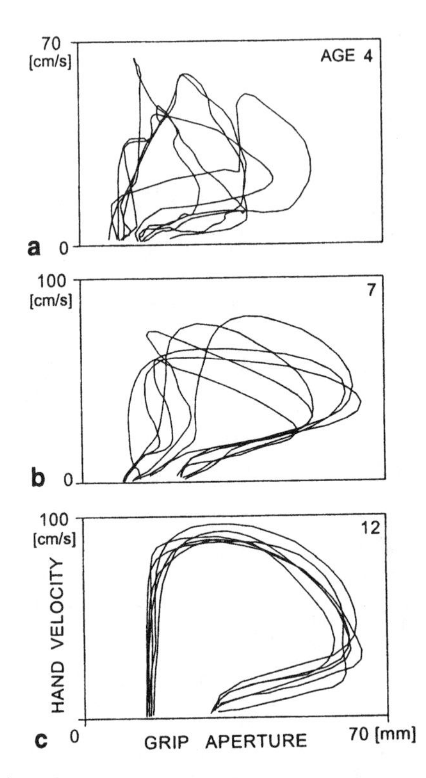

Figure 3–5 Decreased variability during childhood in the coordination of simultaneous movement components (reach and grasp). Kinematic profiles of prehension at the ages of 4, 7, and 12 years. The hand velocity is plotted against the grip aperture in three children of different ages. Six trials are superimposed for each subject. *Source:* Reprinted with permission from J.P. Kuhtz-Buschbeck et al., Development of Prehension Movements in Children: A Kinematic Study, *Experimental Brain Research,* Vol. 122, pp. 424–432, 1998, Springer-Verlag.

hand velocity (on the ordinate) is plotted against grip aperture (on the abscissa) for representative children of different ages (Top: 4 yrs.; Middle: 7 yrs.; Bottom 12 yrs.). As shown in Figure 3–5, the coordinative pattern displayed by the 4-year-old is highly variable, with little timing regularity between the two components. By 7 years of age, some regularity is present. However, it is not until the child reaches 12 years of age that the coordinative pattern starts to become adult-like, that timing of grasping is consistently and smoothly integrated with reaching. Hence, considerable practice, extending over a 12-year span, is required before control processes refine the temporal coupling of these simultaneous movement components.

Consider another example: the coordination of fingertip forces as an object is grasped and lifted. Two phases are discernible in lifting an object using a precision grasp (thumb and index finger): (1) preloading—a firm grasp of the object is established by applying pinch-like forces against the object (grip force) before the vertical (lift) forces are applied; and (2) loading—grip and lift forces increase simultaneously to raise the object from the surface.[102] In adults, the coordination of grip and lift forces during the loading phase are tightly coupled: these forces increase in parallel over time. As demonstrated by Forssberg and colleagues,[103] the coordinative pattern is different for young children. Shown in Figure 3–6 are the grip forces (on the ordinate) plotted against the lift forces (on the abscissa) during the loading phase in three trials for representative children from 1 year to 8 years of age and the pattern for an adult. Note that the youngest children display a negative load force during the initial application of the grip force (they press the object into the supporting surface). When they do apply a positive load force, grip force has already reached a high level. Thus, the two forces are not meshed together in time. In the adult, the parallel application of these two forces is shown by the relatively linear relationship (Figure 3–6). As reported by Forssberg et al.,[103] it is not until adolescence that children consistently demonstrate the tight coupling of grip and lift forces characteristic of the adult pattern. Note the children's high variability in coordinating grip and lift forces during the three trials shown in Figure 3–6. These are typically developing children; the variability that is evident is a normal consequence of learning.

Similar to the variability described with learning to control intersegmental dynamics and the blending of successive components, the variability shown when learning to couple simultaneous movement components represents a search for optimal solutions to the coordinative problem. This search is not mediated by conscious processes. Instead, it appears to reflect exploratory strategies in a self-organized system that yields chance patterns and, eventually, over extended occurrences, funnels interactions toward an efficient coordinative solution.[104]

Summary: Skill Acquisition

Two interdependent learning processes, operating in parallel, are proposed to mediate skill acquisition. Through explicit learning, the performer develops a mapping between his or her morphology and the regulatory environmental conditions leading to increased success in attaining the action-goal. Implicit learning results in the refinement of motor control leading to efficient movement. With practice, explicit processes change at a faster rate than implicit. Thus, during the initial and intermediate phases of learning, explicit processes are most influential. During later phases of learning, implicit processes have the

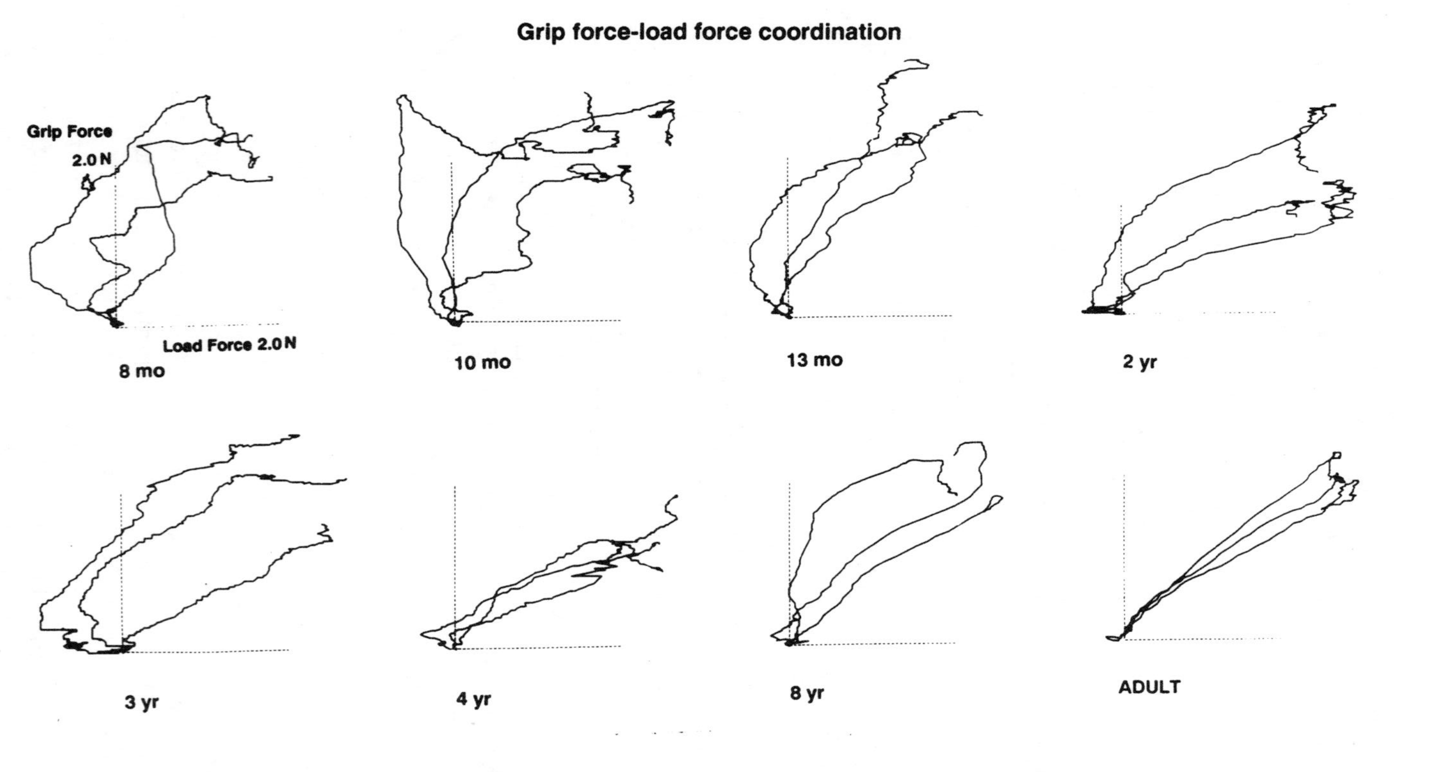

Figure 3–6 Decreased variability from infancy and early childhood to the adult pattern in the coordination of simultaneous movement components while lifting an object. The grip force during the preload and the loading phases is plotted against load forces. Three trials are superimposed for each subject. *Source:* Reprinted with permission from H. Forssberg et al., Development of Human Precision Grip I: Basic Coordination of Force, *Experimental Brain Research*, Vol. 85, pp. 451–457, © 1991, Springer-Verlag.

more dominant role. Let us now consider how therapists can intervene during initial and later phases of learning to facilitate acquisition of skill.

THE THERAPIST'S ROLE IN FACILITATING SKILL

Over the past 40 years, considerable research has been conducted on how skill learning can be enhanced by appropriate teaching strategies. Investigators have examined such training variables as practice schedules, mental rehearsal, augmented feedback, and use of demonstrations. Although these studies have provided a wealth of information, there are several problems in applying the results to learning of functional tasks by patient populations. First, many of these studies have used laboratory tasks involving simple movements that adult subjects already have available in their performance repertoire (eg, simple arm movements required in positioning tasks[105] or barrier knock-down tasks[106]). In a recent and thorough review paper, Wulf and Shea[107] analyzed research on training variables and compared studies using simple laboratory tasks with those using complex, multijoint movements. The results of these studies were so different that Wulf and Shea concluded it is not possible to generalize research findings from simple tasks to learning of complex skills.

Second, even when the task is relatively complex, often it is not a functional activity. In many research situations, the goal (as defined by the experimenter) is to produce a particular movement or to carry out a movement within a specified time interval. Thus, producing a specific movement is the "end" to be achieved. In functional tasks, movement is the "means" to an end; it is the means through which an action-goal is achieved. There are a few situations, usually in sport or dance, where producing a particular movement pattern is necessary. For example, the gymnast or ballet dancer must execute a movement sequence using a prescribed form. However, in coping with one's everyday environment, producing the intended action-outcome is of foremost importance. How that outcome is achieved may be more or less efficient; but, as discussed earlier, many movement patterns can be used to achieve the action-goal. Producing a prescribed movement-form is not necessary to successfully attain a functional goal.

Third, much of the research available has been undertaken using unimpaired adult subjects. Recently, there have been a few reports of training studies in which the subjects were drawn from patient populations.[108–111] This is a promising direction and will provide valuable information needed by therapists as more studies are conducted. Until that time, research based on unimpaired individuals is the primary source of information. However, application of research findings from unimpaired adults to clinical populations has to be made cautiously. In the section that follows, therapeutic implications are drawn from the available research, with concern for generalizability of findings, and with emphasis on studies using complex, functional activities.

Initial Phase of Learning

During initial learning, the performer is trying to discover a movement topology that conforms to environmental constraints and leads to some success in achieving the action-goal. Unimpaired adults typically progress through this phase quite rapidly, sometimes needing only a few attempts at the task. Young children need somewhat more time. However, this initial phase may be much longer for neurological or orthopedic patients who have markedly altered systems, or for elderly persons who may have deficits in explicit short-term memory. In trying to help the patient find a suitable solution to the motor problem, therapists and teachers have only four teaching tools. First, they can give verbal instructions or feedback. Second, supplementary visual input can be provided (such as highlighting of the environment or demonstrating movement patterns). Third, the therapist could manually position the patient or manually guide the patient's movement. Fourth, the therapist can structure the environment for practice. Let us consider how these tools can be used to benefit learning.

Goal Setting

The therapist's first responsibility is to select a functional task, preferably in consultation with the patient, that is potentially achievable by the learner. It is assumed that the patient has been thoroughly evaluated and the task selected is within the patient's capabilities. The second responsibility is to promote the patient's self-efficacy by building the learner's confidence that success is attainable. Patients should be assured that they can achieve the action-goal, although initial attempts may be difficult.

The next responsibility is to avoid goal-confusion. A major difference between learning that occurs naturally and learning that is guided by a therapist or teacher is who establishes the action-goal. When the performer initiates learning, he or she has a clear understanding of the action's purpose. This may not be the case when the goal is set by a teacher or therapist.

Confusion can result when the therapist barely mentions the action-goal and proceeds immediately to a detailed description of the movement. The learner may think that the purpose is to produce the specific movement described by the therapist, rather than achieve some action-outcome. Such confusion is especially evident when practice is arranged under environmental conditions that are not functionally relevant. For example, imagine teaching a patient to reach and grasp a cup by describing the movement and having the performer practice without the cup being present. Consider the same situation but now the instructor starts by focusing on only the reaching component (ie, the therapist has the patient simply move his or her hand to a particular location). In these examples, the goal of the action has been lost and so has the need to map movement to environmental conditions. The trusting learner proceeds through a ritual of training movements (or parts of movements) that are not applicable, without considerable modification, to functional interactions with the environment. Therapists should understand that mimicry of movement is not the same as functional action.

As shown by Wu and colleagues in recent research,[112] stroke patients perform reaching movements more effectively when the environment is set up in a realistic and functional manner. In another study with unimpaired adults, Ferguson and Trombly[113] compared (a) practice involving a functional goal that was carried out in a natural environmental context and (b) practice involving rote exercise. They found enhanced learning when practice was structured in a functional and realistic way. Thus, therapists should set up the task so that it is purposeful and meaningful to the patient. The patient should understand the functional outcome to be produced and the environment should be arranged realistically.

Structuring the Environment

During early learning, all regulatory conditions should be present that are normally operative in the task. For example, when regulatory conditions involve motion in the environment, as in Open or Consistent Motion tasks, such motion must be present during initial practice. The therapist should be aware that practicing first under stationary conditions (a strategy sometimes used by therapists) does not transfer to motion conditions. The two types of tasks pose different requirements. Stationary environments constrain only the movement's spatial organization and do not impose predictive demands. Open tasks impose timing constraints on the movement's organization and require predictions about the object's motion. As was demonstrated recently in a computer-simulated game of table tennis,[114] practice of strokes with a stationary ball does not facilitate performance when hitting a moving ball is required. Learning in these two situations is highly task-specific. Therefore, when Open or Consistent Motion tasks are first practiced, the therapist must arrange the environment to involve the naturally occurring motion of objects, other people, or the terrain.

When tasks involve conditions that typically change from one attempt to the next, as in Variable Motionless or Open tasks, the therapist must decide how to vary the environmental context. For example, suppose the task is reaching to grasp an object. In everyday interactions with the environment, objects vary in size, shape, texture, weight, and—especially—location relative to the performer. The therapist must decide whether successive practice trials should involve the same or different objects that are positioned in the same or different locations.

Over the past 20 years, this issue of practice variability has been examined in many research studies (for reviews see References 115, 116). Contrary to common expectations, the interference produced by randomly changing the environment from one attempt to the next ("random" practice) is often found to benefit learning more than repeated practice under one condition before switching to another ("blocked" practice). Although learning is enhanced following random practice (as measured by retention and transfer tests), performance during practice is better with a blocked arrangement. This paradoxical pattern of findings is referred to as the "contextual interference" effect.

Two proposals have been advanced to account for contextual interference effects. Random practice has been proposed to benefit learning because (a) it causes forgetting of prior plans and thus requires the learner to reconstruct the motor plan on each attempt ("reconstruction" hypothesis[117]); or (b) it promotes the use of diverse strategies in processing information and thus develops elaborate and distinctive memory representations ("elaboration" hypothesis[118]). In both proposals, it is assumed that blocked practice encourages simple repetition of previous information-processing strategies or motor plans without the performer actively engaging in the task. Such repetition may lead to superior performance during practice. However, it does not produce the long-term changes characteristic of learning. As pointed out by Bernstein,[7]

> ...practice, when properly undertaken, does not consist in repeating the *means of solution* of a motor problem time and again, but in the *process of solving* the problem time after time by techniques which we have changed and perfected from repetition to repetition...practice is a particular type of repetition without repetition. (p 134, italics in the original)

Caution is necessary in applying the findings of contextual interference studies to initial learning or relearning in clinical populations. Most research demonstrating the superiority of variable ("random") practice has been conducted with adults using simple, laboratory-type tasks. In a review of contextual interference research, Magill and Hall[115] pointed out that subject variables (age, prior experience, or level of expertise) may determine whether random practice benefits learning. For example, three studies[119–121] in which children served as subjects failed to show the advantage of random over blocked practice. In one of these studies, Del Rey et al.[119] found that blocked practice was superior to random during transfer tests. In another study by Pigott and Shapiro,[120] no differences between random and blocked practice were evident, although a hybrid schedule (serial practice) did produce superior results.

Because children are at an early stage of learning, practice schedules may affect them differently from adults. Supporting this assumption, Farrow and Maschette[122] found that 8- to 9-year-olds profited by blocked practice in learning tennis strokes, whereas random was better for older children (10- to 12-year-olds). Similarly, Pinto-Zipp and Gentile[123] observed that blocked practice of a Frisbee throwing task benefited young children (5- to 6-year-olds), whereas random practice was better for adults. However, when practice was restricted to the very early practice of a Frisbee throw,[123] both adults and children (7- to 9-year-olds) showed a blocked practice advantage. During initial attempts at performing a total body movement like the Frisbee throw, blocked practice may help both children and adults shape a movement topology appropriate for the task. After an appropriate movement form is identified, random practice may be advantageous. This supposition is supported by the research of Shea and colleagues.[124] They demonstrated that acquisition of a force production task was initially enhanced by blocked practice, whereas random practice showed an advantage only during later acquisition. Thus, the random practice benefit typically found in laboratory tasks with adults may indicate that the simple movements tested have already been acquired and that the performer is at an advanced phase of learning.

Blocked practice may be beneficial during initial phases of learning complex (whole-body) tasks. For example, Herbert and colleagues[125] examined learning of tennis strokes by high- and low-skilled performers using blocked or variable practice schedules. Blocked practice enhanced learning for low-skilled performers (assumed to be novices); practice schedules did not differ for high-skilled performers. In contrast to these findings, when the task involves small, manipulative movements (eg, using chopsticks to transport an object[126]), novice performers have been found to benefit more from random practice. These conflicting results may reflect differences in motor control processes required by the tasks. In complex, whole-body movements (such as tennis strokes), the novice's inability to control intersegmental dynamics during early acquisition produces considerable variability in performance. Learning may be disrupted if more variability is added by changing the environmental context (as in random practice). Thus, for complex whole-body tasks, initial learning may be enhanced by blocked conditions. Manipulative tasks, carried out while seated, involve a simpler motor control problem. Performance consistency may be higher than is the case in whole-body movements. Therefore, the contextual variability induced by random practice may be advantageous during initial learning of simple manipulative tasks.

What are the take-home messages for therapists from these contextual interference studies? In arranging practice for Open or Variable Motionless tasks, two generalizations seem reasonable. Blocked practice (limited variability in environmental conditions) may be better than random during the early phases of learning in complex whole-body tasks, especially for individuals who have difficulty controlling intersegmental dynamics (children, novices, or neurological and some orthopedic patients). Random practice may be beneficial during initial learning of simple manipulative tasks or during later learning of complex tasks.

Selective Attention

To effectively plan the movement, the patient must identify and attend to regulatory environmental conditions. In a series of studies using complex tasks (eg, a ski-simulator, balancing on a stabilometer, and golf strokes), Wulf and colleagues[127] have examined how to direct the performer's attentional focus. During initial practice, they found that directing the performer's attention to the goal and to the environmental constraints ("an external focus") leads to better learning than focusing on the movement ("an internal focus"). In some situations, focus on the movement was actually found to be detrimental to learning.

Verbal instructions are the most straightforward way of directing the performer's attention to external events. When a patient's receptive language capabilities are limited (eg, infants, language-delayed children, or aphasic adults), the environment can be arranged to promote an external focus. Regulatory environmental conditions can be highlighted in two ways. First, these environmental features can be accentuated by using color, markings, or contrast with the background. Second, nonregulatory conditions can be eliminated, making important features stand out by reducing distractions. This latter strategy, however, poses some risk. Performance may be disrupted when nonregulatory conditions are reintroduced into the situation.[128]

In Open or Consistent Motion tasks, it is helpful to have the learner first observe the motion patterns of objects, people, or the terrain. As they observe the environment, learners can be encouraged to think about how they will organize their movement's timing. In addition, the therapist can suggest ways to cope with the Open task's predictive demands. The therapist can direct the performer's attention to advance cues that signal upcoming motion patterns of people or objects. Focusing the performer's attention is also useful in Body Transport tasks. Patients can be encouraged to look farther ahead to gather usable information. This is particularly important for neurological or orthopedic patients who must modify their visual scanning strategies to compensate for longer time-lags in their now disabled systems.

Planning the Movement

Instructions about the movement's general form facilitates learning, especially during initial attempts at the task. Demonstrations are useful for this purpose, not only initially, but interspersed throughout practice.[129] There is some truth in the old adage that a picture is worth a thousand words. Thus, seeing someone else perform the task is the most direct way for the learner to understand how to shape movement's topology. The movement's form can be demonstrated by the therapist or—even better—by the patient's peers (ie, other individuals with similar disabilities). It may be beneficial to have several patients show how they perform the task. Demonstrating different approaches accentuates commonalities in the movement's form and de-emphasizes aspects associated with individual performances.

Instruction about details of the movement pattern should be held for later in learning (if given at all). Early in learning, the performer is not able to attend to or remember lengthy verbal descriptions of the movement. Further, it is questionable whether the instructor really knows how to organize detailed patterning of the movement. Although movements in some everyday tasks have been studied (eg, transition from sit-to-stand[130,131]), generally, few functional activities have been analyzed biomechanically. Simple visual observation is not sufficient to derive an understanding of the movement's invariant features. Even skilled observers of movement are sometimes led astray. For years, swimming coaches maintained that the most effective stroke in the crawl required a straight pull back. It was not until Schleihauf[132] carried out several fine-grain analyses of the crawl stroke that an S-shaped pull was found to be most effective. Even if invariant features have been identified, the individual's morphology is still an issue. A movement form is particularized with reference to the structure of the performer. The performer has a better understanding of that structure, altered though it may be through disability, than the assumptions of an instructor. Thus, it would seem best to keep advice about movement organization rather general to start.

During initial attempts, the learner should plan the movement in its entirety. Research has found that approaching the movement as a whole rather than first practicing parts is more beneficial to learning.[133] For example, Winstein and colleagues[134] had hemiparetic patients practice lateral weight shifts while standing. They found no transfer of the "part" practice to the lateral weight shifts required during locomotion. The exception to this guideline of whole-movement practice is a sequential pattern that has natural breaks between components (eg, transfer from wheelchair to toilet preceded by positioning the wheelchair, locking it, and placing the sliding board). In such sequential patterns, part practice (or progressive part practice) may be useful.

Another therapeutic concern is establishing an appropriate speed/accuracy set. In general, performance should be carried out at the same speed and with the same stress on accuracy as required in typical performance. The older notion of first developing accuracy by performing the movement slowly and later speeding up to the required rate, turns out to be in error. Studies have shown that learning is best when practice is undertaken, right from the start, with the customary emphasis on speed and accuracy.[135,136]

The therapist should encourage the learner to imagine the movement before performance and to remember the motor plan afterward. Research has demonstrated that mentally imaging the movement before execution facilitates performance.[137] In addition, instructing the learner to try and remember the motor plan after performance assists the decision-making process that follows performance. Short-term memory of the motor plan is essential for coding information feedback about the movement and making effective decisions about the next response.

Performance of the Movement

During the learner's initial attempts at the task, additional verbal instruction from the therapist could disrupt the patient's performance. The learner is focused on the task: attention is directed toward the environmental constraints and movement organization. Verbal cues or encouragement by the therapist may interfere with performance by stressing the learner's attentional resources. At this point, the important tasks for the instructor are to carefully observe the movement and to safeguard the learner.

Generally, manually guiding the movement is not recommended.[138,139] There are two reasons why guiding the movement may not facilitate learning. First, the task for the learner is to discover how to organize movements that fit the environment. Guiding the movement undermines that problem-solving process. Further, the movement pattern and proprioceptive feedback imposed through manual guidance are different from self-initiated motion. Thus, guidance does not provide an appropriate template for the movement's topology nor does it assist with the motor control problems that the performer must solve. Rather than manual guidance, demonstration can be used to help the learner identify an appropriate movement topology. However, the coordinative problems associated with motor control can only be addressed through the performer's self-initiated movement.

Second, when the therapist places hands on the performer for manual guidance, he or she becomes part of the regulatory environmental conditions. Not only may the performer become dependent on such input, but also when guidance is removed, the environment is changed. The patient must now reorganize his or her movement to fit the altered environmental conditions. Thus, guarding the patient is beneficial, guiding the patient's movement is not.

Augmented Feedback

Information about the learner's performance that supplements intrinsically available information is referred to as augmented feedback. Providing augmented feedback is an important responsibility of the instructor. However, augmented feedback given for informational purposes is different from verbal comments given for motivational reasons. During early learning, performers respond well to encouragement. Keeping the performer "on task" is essential because practice is the most important determinant of learning. Therefore, the instructor can be generous in the use of motivational feedback. Sometimes, however, the instructor's comments are ambiguous. Statements such as "good" or "a fine try" could mean the performance was splendid, or they could mean that the instructor appreciates the patient's effort in attempting the task. When general encouragement is intended, the instructor's verbal statements should make this purpose clear to the performer.

Augmented feedback that provides information to the learner can take two forms: (1) knowledge of results—information feedback about the action-outcome and (2) knowledge of performance—information feedback about the movement. Research studies examining the effects of these two forms of augmented feedback have typically been conducted during early learning. Let us consider the therapeutic implications of this research.

Augmented Feedback about the Outcome. Several research studies using simple tasks have found that intermittent is better than continuous feedback about the action-outcome.[140–142] These findings have been interpreted as supporting the "Guidance hypothesis,"[143,144] which predicts that reducing augmented feedback leads to less error and guides correct performance without the learner becoming overly dependent on this information. In contrast, high availability of feedback supposedly promotes dependency and does not encourage the learner's active involvement in error detection and correction. Thus, augmented knowledge of results has been found to enhance learning when given only occasionally (for example, on 50% of the trials[140]) rather than after every attempt. The availability of knowledge of results has been experimentally reduced in several ways: (a) summary feedback—given after a series of trials rather than after every attempt; (b) average feedback—again given after a series but only specifying the average error on those trials; or (c) bandwidth feedback—providing information only when error exceeds a predefined limit. Using such procedures, studies have shown that decreasing the availability of augmented feedback produces better learning during early practice.[141,142]

A problem exists with the experimental tasks used in many of these studies. As pointed out by Laszlo, the tasks used often lack ecological validity.[145] One common task is lever positioning, carried out without vision (the subject is blindfolded), in which the arm must be moved to experimenter-defined locations or moved within prescribed time periods. Thus, normally available knowledge of results is not present. Such tasks are unlike typical, functional behaviors. Unless the patient is blind, information feedback about the outcome is usually present in everyday activities.

Thus, it is questionable that experimental findings based on these artificial, laboratory-type tasks can be generalized to learning those functional behaviors of interest to therapists. Magill and colleagues[146] examined the effects of adding knowledge of results during learning of a coincidence anticipation task. In their task, intrinsic feedback about the outcome was readily available to the subject. They found that redundantly providing knowledge of results did not increase learning. The comparable therapeutic situations would be telling a stroke patient that he or she failed to successfully transition from sit-to-stand, or fell while trying to locomote from one place to another, or did not grasp an object. Only patients with severe sensory or cognitive impairments would profit from such augmented knowledge of results. For other patients, the information is redundant with intrinsic feedback.

Augmented Feedback About the Movement. During initial practice, providing augmented feedback about the movement (knowledge of performance) may not be beneficial to learning. Movement variability is normally high during this phase of learning as the performer is actively exploring different approaches to the task. Therefore, simply giving information about the prior movement may not be helpful for the next attempt. However, learning is facilitated when knowledge of performance is combined with transitional information that consists of (a) cues to direct attention to key aspects of the movement or (b) information as to how to better organize the movement (error-correcting). In learning an overhand throwing task, Kernodle and Carlton[147] found that combining augmented feedback about the movement with these additional inputs (attention-focusing or error-correcting) was better than providing knowledge of performance or knowledge of results alone. Similarly, Wulf and colleagues[148] demonstrated that learning was enhanced during acquisition of a repetitive slalom-like movement on a ski-simulator when frequent augmented feedback was given about a key feature of the movement.

In another study examining acquisition of an overhand throwing task, Janelle and colleagues[149] varied how often subjects were given knowledge of performance (in combination with transitional inputs: attention-focusing or error-correcting cues). One group received only knowledge of results that was redundant with intrinsic feedback. A second group was given summary information about its movement performance after every five trials. In a third condition (SELF), subjects were provided with knowledge of performance and transitional input whenever they requested it. In a fourth "yoked" condition, other subjects were given the same information on the same schedule as the SELF group. Hence, these "yoked" subjects did not ask for the feedback they received. Although the three groups receiving knowledge of performance improved throwing accuracy, learning was best for subjects in the SELF group that were given information only when they requested it. The implication for therapists is that providing augmented knowledge of performance on a fixed schedule may not be best. It may be better to instruct learners to ask for feedback about the movement whenever they think it would be beneficial.

In dealing with neurological patients, augmented feedback and transitional information might be especially important to offset the use of compensatory movements. Patients who are unable to suitably control motor processes may organize the movement to exploit their intact capabilities. Thus, the action-goal may be attained by substituting available movements for those that are impaired. These compensatory movements may be successful in the short term but may limit functional recovery. Augmented feedback and restructuring the environment can be used to preclude these compensatory responses.

However, the instructor should expect patients to use several maneuvers to make the system more controllable during early practice. Co-contraction of opposing muscle groups to "freeze" degrees of freedom and simplify control is commonly used by unimpaired learners. The same strategy should be expected in patients. As discussed previously, first attempts at organizing a movement do not yield smooth, flowing patterns. Quite the contrary, initial learning is characterized by variable and inefficient movements. Only much later in learning does the performer master the coordinative problems associated with motor control (refinement of intersegmental dynamics, blending of successive components, and coupling of simultaneous movements). Thus, therapeutic concerns about the "quality" of movement are inappropriate during early learning. The patient should be given an opportunity to explore different organizational approaches before the movement is targeted for augmented feedback. The instructor might wait until after the first phase of learning to see if inefficient aspects of the movement are still present. Often, inefficient maneuvers disappear with practice as the learner's active exploration yields more effective approaches.

Delay and Precision of Augmented Feedback. Following performance, augmented feedback about the outcome and the movement should be delayed for a few seconds so the learner can consolidate intrinsically available information. The performer has to organize and code (a) information feedback about the movement with reference to the motor plan and (b) information feedback about the outcome with reference to the action goal. Providing augmented feedback too soon after performance may disrupt these coding processes. The instructor does not have to rush into giving feedback. Research has shown that delaying feedback for a few seconds enhances learning.[150] In addition, augmented feedback should not be too precise during early stages of learning. As was the case in planning the movement, the learner has limited attentional resources and cannot cope with highly detailed information. If necessary to comment on the movement, the instructor should focus on general features (eg, topology) not highly specific components.

Decision Making

To assist the learner in making decisions about the next response, the instructor can provide transitional information, including cues about focusing attention or organizing the movement. If learners are successful in accomplishing the goal and in executing the movement as planned (see Exhibit 3–1), then they should be encouraged to try to reproduce a similar movement on the next attempt. When the movement is executed as planned but the action-goal is not attained, the learner should be guided toward modifying or changing the movement. Repeating a previously unsuccessful movement slows the learning process. Yet, some performers, who have not "learned how to learn," may use this inefficient strategy unless guided not to do so. If the performer continues to be unsuccessful in producing the desired outcome, re-examining the environment may be necessary. The movement may be unsuccessful because the learner failed to notice some critical environmental features. Helping the learner identify relevant conditions is appropriate under these circumstances.

If the performer achieves the goal through an unintended movement, the instructor may be able to describe or demonstrate the pattern that was produced and help the performer reconstruct a similar movement. Therapists are skilled observers and usually can mimic a patient's movement pattern. When an aberrant movement emerges that is successful in reaching the goal, the therapist can serve as an external form of memory by demonstrating the chance pattern.

When the patient fails to move as planned and to reach the goal, therapeutic support is most needed. Without encouragement and additional instruction, learners may apply less effort to the task, alter the action-goal, or quit.

The learner, with the instructor's guidance, should explore movement options until one or more occurrences of total success (movement and goal-attainment). Practice should continue until the performer indicates that he or she has a general concept of an appropriate movement topology. Success in attaining the action-goal is not high at this point. Consistency develops as learning progresses to the next phase.

Later Phases of Learning

During intermediate and later phases of learning, the therapeutic focus is on the learner's development of skill. Recall that skill involves consistent success in attaining the action-goal and economy of effort. Consistency in achieving the goal is determined primarily by explicit learning, whereas efficiency is dependent on implicit learning. Although both processes are influential during all stages of learning, implicit processes have a more dominant effect during later phases. In addition, the nature of the task determines the relative influence of these two learning processes. To facilitate explicit and implicit processes, the instructor has two broad areas of responsibility: (1) structuring practice and (2) providing feedback.

Practice

Learning is directly determined by the amount of practice. Thus, time-on-task is the best predictor of skill acquisition. However, learning rate varies according to how practice is organized. Thus, arranging practice is the most important therapeutic responsibility during later phases of learning.

Structuring Practice: Enhancing Goal Attainment. The environmental context is an important factor in determining how practice should be structured to enhance skill development. Thus, we describe how to arrange practice for tasks within each of the four categories based on environmental context (see Table 3–1). First, however, Closed tasks require special comment. As discussed previously in this chapter, few functional activities in daily life are truly Closed. The patient's prognosis and the therapeutic goals would have to be severely limited for Closed tasks to be the focus of skill development during later learning. Perhaps, the Closed task of putting on one particular pair of shoes using Velcro closures (while sitting on one particular chair) might be a therapeutic objective for an adult with profound mental retardation. However, the therapeutic goal for most patients is to don different types of shoes while sitting on different types of chairs. Similarly, the therapeutic goal rarely is walking on only one type of surface. Even elderly stroke patients who are severely impaired and home-bound have to move about on surfaces that vary (eg, carpet, tile, or wood flooring). Thus, Variable Motionless and Open tasks are more common in the everyday functional activities of most patients.

When acquisition of skill in Closed or Consistent Motion tasks is necessary, regulatory conditions should be held constant during practice. Neurological and orthopedic patients have considerable difficulty controlling movement because of their altered systems. Their movements display high variability even when the environment does not vary. Thus, keeping the environment constant helps the learner to discard crude maneuvers that limit degrees of freedom (eg, freezing joint motion) and to develop some degree of control over intersegmental dynamics and the timing of movement components. By maintaining fixed environmental constraints during practice of Closed and Consistent Motion tasks, the patient is able to refine an appropriate movement pattern, leading to greater success in reaching the goal.

In Open and Variable Motionless tasks, it is crucial to arrange practice so that the environment varies during later phases of learning. In these situations, the learner must diversify the movement to match variations that occur in the normal environment. So, changing conditions from trial-to-trial (ie, random practice) is most beneficial for learning. The instructor may be more comfortable with constant or blocked practice (repeated trials under one condition before switching to the next) because the learner performs better under these limited conditions. However, research on contextual interference demonstrates that performance during practice is not a reliable measure of learning. When retention and transfer tests are given, learning is better following practice under variable than under constant or blocked conditions. Therefore, in Open and Variable Motionless tasks, it is the instructor's responsibility to systematically change the regulatory conditions during practice to facilitate skill acquisition.

Structuring the Environment: Enhancing Efficiency. Variability in practice conditions also promotes movement efficiency by increasing the likelihood that better solutions to the coordinative problem will emerge. However, the key factor is the amount of practice. Implicit learning processes mediating movement efficiency change slowly as a consequence of prolonged practice. Everyday functional activities are practiced continuously over the life span by unimpaired individuals. As we walk, climb stairs, reach for objects, move from sit-to-stand, and perform other common activities, efficiency develops as a natural consequence of ongoing engagement in our daily activities. In nonessential tasks (eg, playing a musical instrument, sport, or dance activities), smooth efficient movement emerges only when the performer is sufficiently motivated to practice for an extended period of time. Thus, facilitating implicit learning leading to movement efficiency requires practice, practice, and more practice.

Following neurological damage in adults, efficient movement is often disrupted. For example, stroke patients who have sustained damage to cortical motor areas interrupting projections to or directly involving the basal ganglia have persistent motor impairments, particularly in the regulation of intersegmental dynamics.[151,152] In contrast, such damage seems to spare the representational processes associated with the movement's topology.[151] In visually guided performance, these patients are able to spatially map movement to the environment. In relearning functional tasks acquired before their cerebral vascular accident (CVA), they already have a general concept of how to shape the movement's topology to match environmental constraints. Thus, these stroke patients are beyond the initial stage of learning. Their problem is in controlling processes underlying effective coordination.

Research by Platz and colleagues[151] demonstrates this pattern of sparing and impairment. They examined learning in patients who had almost completely recovered from hemiparesis following a single CVA. Almost all patients had damage involving the basal ganglia. They used a spatial-motor task that required moving the upper limb and hand in a triangular pattern within a confined space. Kinematic analysis of movement indicated that these patients were able to solve the spatial-motor problem: they displayed an appropriate movement topology. However, their movement patterns were highly irregular, indicating problems in coordination and control. In comparison with unimpaired subjects, patients' movements were characterized by (a) longer and more variable movement times, (b) higher variability in several other movement parameters (eg, velocity and mean square jerk), and (c) many more submovements and segmentation (number of zero-crossings in the linear acceleration profile). The investigators[151] concluded that damage to motor cortex and the cortical-basal ganglion loop produced a deficit in "automatic control processes" that, in the present analysis, is interpreted as an impairment of implicit learning.

In another study, Levin[152] produced similar findings in pointing movements of hemiparetic patients. Again, most patients had involvement of the basal ganglia. The task required the patient to slide his or her arm along a table to contact targets (in near or far positions) that were located straight ahead of, ipsilaterally, or contralaterally to the starting position. Levin found that patients could spatially organize movement to attain all targets, indicating no impairment in mapping movements to the environment. However, interjoint coordination (an "automatic control process") was impaired.

How can the therapist facilitate recovery of these motor control processes that are not consciously available? Within the limits of plasticity permitted by the damaged system, reorganization can occur only through the patient's active practice of diverse movement patterns. However, the patient has difficulty effectively moving the paretic limb. Without therapeutic intervention, patients tend not to use the limb (ie, learned disuse[153]) or to limit movement options to those yielding most control. The therapeutic goal with such patients is to encourage use of the impaired limb and to promote movement variability so that implicit relearning and functional recovery can occur.

Research has shown that two ways of arranging practice can assist stroke patients with motor control problems.[153,154] The first approach is referred to as "forced use" or "constraint-induced movement" (CIM).[153,155] The unimpaired upper extremity of a chronic stroke patient is placed in a sling to prevent it from being used in daily activities. In a study by Wolf and colleagues,[153] patients wore the sling during waking hours for 2 weeks (except for 30 minutes per day). The therapeutic object was to overcome the patient's learned nonuse by requiring mobilization of the impaired limb in functional activities. Patients demonstrated improvement immediately after the intervention and sustained this improvement in a 1-year follow-up test for 19 of the 21 functional tasks. Taub and Wolf[155] have reported studies using CIM that replicate and extend these prior findings. Further, "forced use" was recently applied in the therapy of a 2-year-old child with cerebral palsy and beneficial effects were also found.[156] However, systematic evaluation of forced use in children has not been carried out. Therapists should be wary of applying this intervention to young children until more is known about its effects on all aspects of development (cognitive, social, and motoric). In adults, studies of forced use or CIM seem to indicate that compelling use of the impaired limb in diverse functional activities enables stroke patients with motor control impairments to regain some functional capabilities.

A second way of arranging practice is referred to by Dean and Shepherd[154] as task-related training. Using this approach, they have experimentally shown improved performance in stroke patients. In their study, seated patients reached and grasped objects with their unimpaired limb. Thus, the patient's primary goal was to acquire the object. However, the therapeutic goal was to increase support by the impaired lower limb: (a) to enable reaching farther and faster and (b) to promote use of the impaired lower limb in other functional activities. In arranging practice, the position and distance of objects were varied to require better lower-limb support by the patient. In addition, the patient was verbally encouraged to increase reaching speed, which further challenged use of the impaired lower limb.

In Dean and Shepherd's reaching tasks, effective lower-limb support necessitated more than the production of active muscle force.[154] During reaching, their patients had to control and exploit the passively induced linkage effects that required feedforward processes. As discussed previously, motion at one joint sets up mechanical interaction forces that radiate through the multi-linked kinematic chain, passively producing torques at remote joints. These intersegmental effects resulting from the reaching movement (especially when executed rapidly) had to be accounted for in advance so to preserve body stability and the hand's trajectory.

The stroke patients in Dean and Shepherd's study were examined before and after 2 weeks of training on (a) weight bearing in standing and during a sit-to-stand transition using force-plate measures and (b) reach performance using detailed movement analysis.[154] They found that "task-related training" enabled patients to reach further and faster indicating improved support by the impaired lower limb. In addition, improved weight bearing was found in the sit-to-stand transition. Apparently, this training protocol promoted feedforward processes that enhanced performance in the practiced task (reaching) and produced positive transfer to the sit-to-stand task, perhaps because of similarities between the two tasks in the control of intersegmental dynamics.

Dean and Shepherd's[154] use of the term "task-related" could be confusing. In terms of the present analysis of learning, their patients were consciously trying to attain the functional goal of reaching to grasp an object. This aspect of task-performance was mediated by explicit processes. However, weight bearing by the impaired lower limb during reaching implicated implicit processes. Therefore, this task-related approach to therapy emphasized the functional action and not the consciously unavailable process of controlling intersegmental force dynamics. By systematically varying environmental conditions governing the functional action, effective weight bearing was indirectly elicited. Thus, the improved use of the lower extremity unconsciously "fell-out" of the goal-oriented reaching task: it resulted from implicit learning.

In extending this therapeutic approach to other tasks, the challenge for therapists is to creatively design ways of structuring practice that will elicit feedforward control of intersegmental dynamics within the context of goal-directed, functional activity. The therapist must devise ways of systematically varying environmental conditions to indirectly compel use of the impaired control processes so that implicit learning (or relearning) can occur.

Scheduling Practice. The therapist must decide whether practice trials should be clustered together (massed) or spaced over time (distributed). In general, performance but not learning is better under distributed than massed practice. For example, Durham[157] observed that the apparent superiority of distributed over massed practice disappeared following a brief rest period when experimental groups were tested under the same spaced conditions. Under these conditions, performance of the group originally trained under massed conditions jumped to a level comparable with the distributed group. Massing may produce poorer performance because there is little opportunity to rest between attempts. During closely spaced trials, fatigue or other negative states may depress the level of goal-attainment. However, learning still occurs.

The endurance of patients may be limited especially in difficult tasks. Learning continues even though performance may be depressed by fatigue. After a rest period, an abrupt improvement is often evident (technically termed "reminiscence"). However, performance decrements resulting from fatigue may discourage patients. The instructor should try to sustain the patient's motivation during massed practice. It might be useful to explain to the patient that learning is progressing even though performance is degraded.

Mental Practice. In arranging practice, the last factor to be considered is the efficacy of mental rehearsal. Research has shown that mental practice improves performance in many tasks.[158–160] It is most effective after the individual has progressed beyond the first phase of learning and has developed a basic understanding of the task's constraints and the movement's topology. Further, Closed and Variable Motionless tasks seem to profit more than tasks involving motion in the environment. When mental practice does enhance learning, it is not as potent as physical practice. Consciously accessing the movement's topology and the environmental constraints probably contributes to explicit learning, while having little (if any) benefit for implicit learning. Thus, the most productive arrangement seems to be a combination of physical practice and mental rehearsal.

Time in therapy is limited and, probably, best spent on physical practice of the task. However, patients can be instructed to rehearse the task mentally when they are not in therapy. From a meta-analysis of research findings, mental practice was found to be optimal in unimpaired adults when limited to 20-minute bouts.[161] Similar information regarding the optimal practice period is not available for clinical populations. However, in advising patients about mental rehearsal, brief practice bouts (20 minutes or less) would seem better than longer ones.

Augmented Feedback

During later learning, instructors can provide information feedback about the movement (knowledge of performance) or information feedback about the performer-environment interaction (knowledge of results). In both cases, feedback from the instructor is directed toward enhancing explicit learning processes. It does not seem likely that augmented feedback would facilitate implicit learning processes (eg, motor control associated with coupling of grip and load force during lifting, or exploiting motion-dependent forces during the swing phase in walking). Voluntary regulation of these motor control processes does not appear possible.

To enhance explicit learning, the task determines which type of augmented feedback is most appropriate later in practice. When the instructor is trying to help the performer develop a relatively stereotyped pattern of movement (as in Closed and Consistent Motion tasks), it is advantageous to give augmented feedback about the movement. Given that environmental conditions remain the same in these tasks, learners strive for consistent reproduction of an effective movement pattern. Augmented feedback about the prior movement in combination with transitional information aids the learner in correcting or refining this pattern.

Providing information about the movement seems best given after the performance (terminal feedback) rather than during the movement (concurrent feedback). For example, Engardt[162] gave extensive training to chronic stroke patients using concurrent auditory feedback to signal maintenance of symmetrical body weight distribution while standing up and sitting down. Control subjects (also stroke patients) simply practiced without augmented feedback. Immediately after 6 weeks of daily training, significant improvement was evident in standing up for both the feedback and control groups, whereas only the feedback group showed improvement in sitting down. However, 33 months later, groups did not differ on either transition. It appears that the amount of practice, not augmented feedback, was the critical element for both groups. During training, the concurrent augmented feedback did produce improvement in weight distribution while sitting down. However, this feedback was used as a "crutch." No long-term, carryover effect occurred when it was removed. Similar results have been found in studies using biofeedback. In an extensive review and meta-analysis of research findings, Moreland and

Thomson[163] concluded that providing augmented biofeedback of muscle contraction patterns has no more benefit than standard therapy.

In Open and Variable Motionless tasks, environmental conditions change from one trial to the next. Augmented feedback about the movement used on the last attempt is not beneficial because a different movement organization may be required on the next try. Unlike performance in Closed and Consistent Motion tasks, organization of the movement is not based on reproductive memory. Open and Variable Motionless tasks require ongoing problem solving. The learner must generate new movement compositions that fit the variable environmental conditions. In these tasks, feedback about prior movements is less important than providing cues that emphasize the more general and invariant features of an effective pattern.[148] Such cueing can be used to help the performer precisely shape movement to environmental constraints.

In addition, performers need augmented feedback about the fit between their assessment of the environment and the movement's organization. Was the general approach appropriate for these conditions? Did the movement's form match the environmental context? These are the relevant domains that instructors can address. Information-processing demands distinguish Open from Variable Motionless tasks. Learners need augmented feedback about the adequacy and rapidity of their predictions in Open tasks. Thus, therapists must be more than good diagnosticians of movement. They must be able to analyze how well the learner is coping with all the challenges posed by the task. Along with the performer, therapists must monitor environmental events controlling the action. Further, the therapists have the added burden of observing the performer's behavior in this complex setting. Augmented feedback should indicate how precise was the performer's prediction of environmental events and how adeptly was the movement organized. In Open tasks, it is difficult to capture the flow of events in the environment and the performer's response by using verbal feedback alone. Frequently, athletic coaches re-create the performer-environment interaction by setting up similar conditions and replaying the various exchanges that occurred. Therapists might consider this more dynamic procedure for giving augmented feedback about performer-environment interactions.

Beyond these guidelines, two other suggestions concerning augmented feedback can be offered. During skill acquisition, higher-level units for coordinating movements and for processing environmental input are developed. Thus, attentional demands during and after task-performance are reduced. Now, the performer can attend to more detailed information from the therapist. Therefore, precision of augmented feedback can be increased. Second, augmented feedback does not need to follow every attempt at the task. During later phases of learning, frequency of augmented feedback should be reduced to encourage more independent learning by the performer.[145]

FINAL SUMMARY: FACILITATION OF SKILL ACQUISITION

The processes underlying acquisition of functional skill depend on the task and the phase of learning. Skill development is enhanced by appropriate instruction. During initial practice, the therapist helps the performer identify task-constraints and an appropriate movement topology. Later in learning, the therapeutic focus is on increasing success in attaining the action-goal and developing movement efficiency. During both phases of learning, structuring the environment and arranging practice conditions are important responsibilities of the instructor. In addition, augmented feedback can be given along with transitional information to help the learner focus on key aspects of the movement and the environment. However, providing redundant knowledge of results is unnecessary as it is normally available through intrinsic feedback. Finally, no rigidly defined teaching techniques can be applied to enhance skill learning. Therapeutic interventions must be creatively designed based on the characteristics of the performer, the task, and the phase of practice. Thus, effective therapy must complement the learner and learning processes.

REFERENCES

1. Gentile AM. Implicit and explicit processes during acquisition of functional skills. *Scand J Occup Ther.* 1998;5:7–16.
2. Silver M. Purposive behavior in psychology, and philosophy: a history. In: Frese M, Sabini J, eds. *Goal Directed Behavior: The Concept of Action in Psychology.* Hillsdale, NJ: Lawrence Erlbaum Associates; 1985:3–19.
3. Cranach M von. The psychological study of goal-directed action: basic issues. In: Cranach M von, Harre R, eds. *The Analysis of Action.* Cambridge, England: Cambridge University Press; 1982.
4. Flanagan JR, Ostry DJ, Feldman AG. Control of human jaw and multi-joint arm movement. In: Hammond GE, ed. *Cerebral Control of Speech and Limb Movements.* Amsterdam: North-Holland; 1990:29–58.
5. Haaland KY, Harrington DL. Complex movement behavior: towards understanding cortical and subcortical interactions regulating control processes. In: Hammond GE, ed. *Cerebral Control of Speech and Limb Movements.* Amsterdam: North-Holland; 1990:169–200.
6. Gentile AM. A working model of skill acquisition with applications to teaching. *Quest.* 1972;17:3–23.
7. Bernstein A. *The Coordination and Regulation of Movement.* New York: Pergamon Press; 1967.
8. Morasso P, Sanguineti V. Neural models of distributed motor control. In: Stelmach GE, Requin J et al., eds. *Tutorials in Motor Behavior, 2.* Amsterdam: North-Holland; 1992:3–30.
9. Arbib MA. Perceptual structures and distributed motor control. In: Brooks VB, ed. *Handbook of Physiology Vol. II: Motor Control.* Baltimore: Williams and Wilkins; 1981:1449–1480.
10. Abbs JH, Cole KJ. Neural mechanisms of motor equivalence and good achievement. In: Wise SP, Evarts EV, eds. *Higher Brain Functions.* New York: John Wiley & Sons; 1987:15–43.
11. Gentile AM, Higgins JI, Miller EA, et al. Structure of motor tasks. In: *Mouvement, Actes du 7 Symposium en Apprentissage Psycho-motor du Sport.* Quebec, Canada: Professionalle de L'Activite Physique du Quebec; 1975:11–28.

12. Iberall T, Bingham C, Arbib MA. Opposition space as a structuring concept for the analysis of skilled hand movements. In: Heuer H, Fromm C, eds. *Generation and Modulation of Action Patterns, Exp Brain Res.* 1986;15:158–173.
13. Iberall T, Arbib MA. Schemas for the control of hand movements: an essay on cortical localization. In: Goodalle MA, ed. *Vision and Action. The Control of Grasping.* Norwood, NJ: Ablex; 1990:204–242.
14. Greenlee MW, Lang HJ, Mergner T, et al. Visual short-term memory of stimulus velocity in patients with unilateral posterior brain damage. *J Neurosci.* 1995;15:2287–2300.
15. Elliott D. Intermittent versus continuous control of manual aiming movements. In: Proteau L, Elliott D, eds. *Vision and Motor Control.* Amsterdam: North-Holland; 1992:33–48.
16. Allport DA, Antonis B, Reynolds P. On the division of attention: a disproof of the single channel hypothesis. *Q J Exp Psychol.* 1972;24:225–235.
17. Proteau L, Marteniuk RG, Levesque L. A sensorimotor basis for motor learning: evidence indicating specificity of practice. *Q J Exp Psychol.* 1992;44A:557–575.
18. Proteau L. On the specificity of learning and the role of information for movement control. In: Proteau L, Elliott D, eds. *Vision and Motor Control.* Amsterdam: North Holland; 1992:65–101.
19. Broderick MP, Newell KM. Coordination patterns in ball bouncing as a function of skill. *J Motor Behav.* 1999;31:165–188.
20. Darling WG, Cooke JD. Changes in the variability of movement trajectories with practice. *J Motor Behav.* 1987;19:291–309.
21. Ludwig DA. EMG changes during acquisition of a motor skill. *Am J Phys Med.* 1982;61:229–243.
22. Hobart DJ, Vorro JR, Dotson CO. Synchronized myoelectric and cinematographic analysis of skill acquisition. *J Hum Movt Stud.* 1978;4:155–166.
23. Beggs WDA, Howarth CI. The movement of the hand towards a target. *Q J Exp Psychol.* 1972;24:448–453.
24. Schmidt RA. A schema theory of discrete motor skill learning. *Psychol Rev.* 1975;82:225–260.
25. Keele SW. Learning and control of coordinated motor programs: the programming perspective. In: Kelso JAS, ed. *Human Motor Behavior.* Hillsdale, NJ: Lawrence Erlbaum Associates; 1982:161–186.
26. van Ingen Schenau GJ, van Soest AJ, Gabreels JM, et al. The control of multi-joint movements relies on detailed internal representations. *Hum Movt Sci.* 1995;14:511–538.
27. Young RP, Marteniuk RG. Stereotypic muscle-torque patterns are systematically adopted during acquisition of a multi-articular kicking task. *J Biomech.* 1998;31:809–816.
28. Higgins JR, Spaeth RK. Relationship between consistency of movement and environmental condition. *Quest.* 1972;17:61–69.
29. Shiffrin RM, Schneider W. Controlled and automatic human information processing: II. Perceptual learning, automatic attending, and a general theory. *Psychol Rev.* 1977;84:127–190.
30. Cuqlock-Knopp VG, Wilkins CA, Torgerson WS. Multiple cue probability learning and the design of information displays for multiple tasks. In: Damos DL, ed. *Multiple-Task Performance.* London: Taylor & Francis; 1991:139–152.
31. Gibson JJ. *The Ecological Approach to Visual Perception.* Hillsdale, NJ: Lawrence Erlbaum Associates; 1986.
32. Jeka JJ, Lackner JR. Fingertip contact influences human postural control. *Exp Brain Res.* 1994;100:495–502.
33. Kaminski T, Bock C, Gentile AM. The coordination between trunk and arm motion during pointing movements. *Exp Brain Res.* 1995;106:457–466.
34. Bouisset S, Zattara M. A sequence of postural movement precedes voluntary movement. *Neurosci Lett.* 1981;22:263–270.
35. Massion J. Postural changes accompanying voluntary movements: normal and pathological aspects. *Hum Neurobiol.* 1984;2:261–267.
36. Gopher D. The skill of attention control: acquisition and execution of attention strategies. In: Meyer DE, Kornblum S, eds. *Attention and Performance XIV.* Cambridge, MA: MIT Press; 1993:299–322.

37. McCaffrey-Easley A. Dynamic assessment for infants and toddlers: the relationship between assessment and the environment. *Pediatr Phys Ther.* 1996;8:62–69.
38. Branton P. The train driver. In: Singleton WT, ed. *The Analysis of Practical Tasks.* Baltimore: University Park Press; 1978:169–188.
39. Bartlett FC. *Thinking: An Experimental and Social Study.* London: Allen and Unwin; 1958.
40. Seger CA. Implicit learning. *Psychol Rev.* 1994;115:163–196.
41. Snoddy GS. Learning a stability. *J Appl Psychol.* 1926;10:1–36.
42. Crossman ERFW. A theory of the acquisition of speed-skill. *Ergonomics.* 1959;2:153–166.
43. Logan GD. Towards an instance theory of automatization. *Psychol Rev.* 1988;95:492–527.
44. MacKay DG. The problem of flexibility and fluency in skilled behavior. *Psychol Rev.* 1982;89:483–506.
45. Newell A, Rosenbloom PS. Mechanism of skill acquisition and the law of practice. In: Anderson JR, ed. *Cognitive Skills and Their Acquisition.* Hillsdale, NJ: Lawrence Erlbaum Associates; 1981:1–55.
46. Willingham DB. Systems of motor skill. In: Squire LA, Butters N, eds. *Neuropsychology of Memory.* 2nd ed. New York: Guilford Press; 1992:166–178.
47. Jordan MI. Computational aspects of motor control and motor learning. In: Heuer H, Keele SW, eds. *Handbook of Perception and Action. vol. 2: Motor Skills.* London: Academic Press; 1996:71–120.
48. Sparrow WA, Newell KM. Metabolic energy expenditure and the regulation of movement economy. *Psychonomic Bull Rev.* 1998;5:173–196.
49. Fitts PM. Factors in complex skill learning. In: Glaser R, ed. *Training, Research and Education.* Pittsburgh, PA: University of Pittsburgh Press; 1962:177–197.
50. Sparrow WA, Irizarry-Lopez VM. Mechanical efficiency and metabolic cost as measures of learning a novel gross motor task. *J Motor Behav.* 1987;19:240–264.
51. Neisser U. From direct perception to conceptual structure. In: Neisser U, ed. *Concepts and Conceptual Development: Ecological and Intellectual Factors in Development.* New York: Cambridge University Press; 1987:11–24.
52. Reason JT. Actions not as planned. In: Underwood G, Stevens R, eds. *Aspects of Consciousness.* London: Academic Press; 1979:67–89.
53. Brooks VB. *The Basis of Motor Control Systems.* New York: Oxford University Press; 1986.
54. Viviani P, Terzuolo C. Space-time invariance in learned motor skills. In: Stelmach GE, Requin J, eds. *Tutorials in Motor Behavior.* Amsterdam: North-Holland; 1980: 525–533.
55. Goodmay D, Kelso JAS. Are movements prepared in parts? Not under compatible (naturalized) conditions. *J Exp Psychol: Gen.* 1980;109:475–495.
56. Turvey MT, Fitch HL, Tuller B. The Bernstein perspective: 1. The problems of degrees of freedom and context-conditioned variability. In: Kelso JAS, ed. *Human Motor Behavior.* Hillsdale, NJ: Lawrence Erlbaum Associates; 1982:271–282.
57. Vereijken B, Van Emmerik RE, Whiting HT, et al. Free(z)ing degrees of freedom in skill acquisition. *J Motor Behav.* 1992;24:133–142.
58. Gelfand IM, Gurfinkel VS, Tsetlin ML, et al. Some problems in the analysis of movements. In: Gelfand IM, Gurfinkel VS, Fomin SV, et al., eds. *Models of the Structural-Functional Organization of Certain Biological Systems.* Cambridge, MA: MIT Press; 1971:329–345.
59. Delcomyn F. Neural basis of rhythmical behavior in animals. *Science.* 1980;210:492–498.
60. Grillner S, Orlovski G. Locomotion, neural networks. *Encyclopedia of Human Biology, Vol. 4.* New York: Academic Press; 1991:769–781.
61. Greene PH. Problems of organization of motor systems. In: Rosen IR, Snell FM, eds. *Progress in Theoretical Biology, Vol. 2.* New York: Academic Press; 1972:303–338.
62. Kelso JAS, Ding M. Fluctuations, intermittency, and controllable chaos in biological coordination. In: Newell KM, Corcos DM, eds. *Variability and Motor Control.* Champaign, IL: Human Kinetics; 1993:291–316.

63. Skinner R. *The Decision Processes during Acquisition of a Complex Skill.* New York: Teachers College, Columbia University; 1974. Dissertation.
64. Young RP, Marteniuk RG. Changes in the inter-joint relationships of muscle moments and powers accompanying the acquisition of a multi-articular kicking task. *J Biomech.* 1995;28:701–713.
65. Gentilucci M, Daprati E, Toni I, et al. Unconscious updating of grasp motor program. *Exp Brain Res.* 1995;105:291–305.
66. Cohen A, Ivry RI, Keele SW. Attention and structure in sequence learning. *J Exp Psychol: Learn, Mem, Cogn.* 1990;16:17–30.
67. Wulf G, Schmidt RA. Variability of practice and implicit motor learning. *J Exp Psychol: Learn, Mem, Cogn.* 1997;23:987–1006.
68. Magill RA. Knowledge is more than we can talk about: implicit learning in motor skill acquisition. *Res Q Exer Sport.* 1998;69:104–110.
69. McIntyre J, Berthoz A, Lacquaniti F. Reference frames and internal models for visuo-manual coordination: what can we learn from microgravity experiments? *Brain Res Rev.* 1998;28:143–154.
70. McKinley PA. Planning of motor sequences in landing from a jump-down. In: Stelmach GE, Requin J, eds. *Tutorials in Motor Behavior, 2.* Amsterdam: North-Holland; 1992:713–723.
71. Bouisset S, Zatara M. Biomechanical study of the programming of anticipatory postural adjustments associated with voluntary movement. *J Biomech.* 1987;20:735–742.
72. Friedli WG, Cohen L, Hallett M, et al. Postural adjustments associated with rapid voluntary arm movement: biomechanical analysis. *J Neurol Neurosurg Psychiatry.* 1988;51:232–243.
73. Lee WA. Anticipatory control of postural and task muscles during rapid arm flexion. *J Motor Behav.* 1980;12:185–196.
74. Mena D, Mansour JM, Simon SR. Analysis and synthesis of human swing leg motion during gait and its clinical applications. *J Biomech.* 1981;14:823–832.
75. Zernicke RF, Schneider K, Buford JA. Intersegmental dynamics during gait: implications for control. In: Patla AE, ed. *Adaptability of Human Gait.* Amsterdam: Elsevier Science; 1991:187–202.
76. Patla AE, Prentice, SD. The role of active forces and intersegmental dynamics in the control of limb trajectory over obstacles during locomotion in humans. *Exp Brain Res.* 1995;106:499–504.
77. Cionni G, Ferrari F, Prechtl HFR. Motor assessment in the neonatal period. In: Fedrizzi E, Avanzini G, Crenna P, eds. *Motor Development in Children.* London: John Liberty; 1994:13–24.
78. Thelan E. Motor development: a new synthesis. *Am Psychol.* 1995;50:79–95.
79. von Hofsten C. Development of visually directed reaching: the approach phase. *J Hum Movt Stud.* 1979;5:160–178.
80. Konczak J, Borutta M, Topka H, et al. The development of goal-directed reaching in infants: hand trajectory formation and joint torque control. *Exp Brain Res.* 1995;106:156–168.
81. Konczak J, Dichgans J. The development towards stereotypic arm kinematics during reaching in the first 3 years of life. *Exp Brain Res.* 1997;117:346–354.
82. Gilchrist L, Craib M, Morgan D. Variability in the treadmill walking patterns of children, Abstract. *Proceedings: Annual Conference on Gait and Locomotion.* 1997:145–146.
83. Law S-H, Gentile AM, Bassile CC. Gait adaptations of children and adults while stepping over obstacles. *Soc Neurosci Abstracts.* 1996;22:2038.
84. Pilchik-Kolder E, Gentile AM. Ascending a step: age-related changes in movement variability of young children, Abstract. *Proceedings: Neurological Section.* Seattle, WA: American Physical Therapy Association; 1999.
85. Clark JE, Phillips SJ, Peterson R. Developmental stability in jumping. *Dev Psychol.* 1989;25:929–935.
86. Guarrera-Bowlby P, Gentile AM, Nills B, et al. The effect of task-related training on the sit-to-stand transition in children with and without cerebral palsy, Abstract. *Proceedings: Annual Conference.* New Jersey: American Physical Therapy Association; 1997.

87. McDonald PV, Vanemmerik REA, Newell KM. Effect of task constraints on limb kinematics in a throwing task. *J Mot Behav.* 1989;21:245–264.
88. Moore SP, Marteniuk RG. Kinematic and electromyographic changes that occur as a function of learning a time-constrained aiming task. *J Motor Behav.* 1986;4:397–426.
89. Schneider K, Zernicke, RF, Schmidt RA, Hart TJ. Changes in limb dynamics during the practice of rapid arm movements. *J Biomech.* 1989;22:805–817.
90. Schneider K, Zernicke RF. Jerk-cost modulations during the practice of rapid arm movements. *Biol Cybern.* 1989;60:221–230.
91. Ivry R. Representational issues in motor learning: phenomena and theory. In: Heuer H, Keele SW, eds. *Handbook of Perception and Action. Vol. 2: Motor Skills.* London: Academic Press; 1996:263–330.
92. Rumelhart DE, Norman DA. Simulating a skilled typist: a study of skilled cognitive-motor performance. *Cogn Sci.* 1982;6:1–6.
93. Johnels B, Ingvasson P, Holmberg B, Matousek M, Steg G. Single-dose-L-dopa response in early Parkinson's disease: measurement with optoelectronic recording techniques. *Movt Disord.* 1993;8:56–62.
94. Rosenbaum DA, Vaughan J, Jorgensen MJ, et al. Plans for object manipulation. In: Meyer DE, Kornblum S, eds. *Attention and Performance, vol. XIV.* Cambridge, MA: MIT Press; 1992:803–820.
95. Jakobson LS, Goodale MA. Factors affecting higher-order movement planning: a kinematic study of human prehension. *Exp Brain Res.* 1991;86:199–208.
96. Chieffe S, Gentilucci M. Coordination between the transport and the grasp components during prehension movements. *Exp Brain Res.* 1993;94:471–477.
97. Marteniuk RG, MacKenzie CL, Jeannerod M, et al. Constraints on human arm movement trajectories. *Can J Psychol.* 1987;41:365–378.
98. Jeannerod M, Arbib MA, Rizzolatti G, et al. Grasping objects: the cortical mechanisms of visuomotor transformations. *TINS.* 1995;18:314–320.
99. Hoff B, Arbib MA. Models of trajectory formation and temporal interaction of reach and grasp. *J Motor Behav.* 1993;25:175–193.
100. Jeannerod M. Reaching and grasping: parallel specification of visuomotor channels. In: Heuer H, Keele SW, eds. *Handbook of Perception and Action. Vol. 2, Motor Skills.* London: Academic Press; 1996:405–502.
101. Kuhtz-Buschbeck JP, Stolze H, Johnk K, et al. Development of prehension movements in children: a kinematic study. *Exp Brain Res.* 1998;122:424–432.
102. Gordon AM, Forssberg H. Development of neural mechanism underlying grasping in children. In: Connolly KJ, Forssberg H, eds. *Neurophysiology & Neuropsychology of Motor Development.* London: Mac Keith Press; 1997:214–231.
103. Forssberg H, Eliasson AC, Kinoshita H, et al. Development of human precision grip I: basic coordination of force. *Exp Brain Res.* 1991;85:451–457.
104. Sporns O, Edelman GM. Solving Bernstein's problem: a proposal for the development of coordinated movement by selection. *Child Dev.* 1993;64:960–981.
105. Guay M, Salmoni A, Lajoie Y. Summary knowledge of results and task processing load. *Res Q Exerc Sports.* 1997;68:167–171.
106. Shea JB, Morgan RL. Contextual interference effects on the acquisition, retention, and transfer of a motor skill. *J Exp Psychol: Hum Learn Mem.* 1979;5:179–187.
107. Wulf G, Shea CH. Principles derived from study of simple skills do not generalize to complex skill learning. *Psychol Rev.* In press.
108. Hanlon RE. Motor learning following unilateral stroke. *Arch Phys Med Rehabil.* 1996;77:811–815.
109. Dick MB, Shankle RW, Beth RE, et al. Acquisition and long-term retention of a gross motor skill in Alzheimer's disease patients under constant and varied practice conditions. *J Gerontol.* 1996;51B:103–111.
110. Winstein CJ, Merians AS, Sullivan KJ. Motor learning after unilateral brain damage. *Neuropsychologia.* 1999;37:975–987.

111. Shumway-Cook A, Anson D, Haller S. Postural sway biofeedback: its effect on reestablishing stance stability in hemiplegic patients. *Arch Phys Med Rehabil.* 1988;69:395–400.
112. Wu C-Y, Trombly SC, Lin K, et al. Effects of object affordances on reaching performance in persons with and without cerebrovascular accident. *Am J Occup Ther.* 1998;52:447–456.
113. Ferguson JM, Trombly CA. The effect of added-purpose and meaningful occupation on motor learning. *Am J Occup Ther.* 1997;51:508–515.
114. Todorov W, Shadmehr R, Bizzi E. Augmented feedback presented in a virtual environment accelerates learning a difficult motor task. *J Mot Behav.* 1997;29:147–158.
115. Magill RA, Hall KG. A review of the contextual interference effect in motor skill acquisition. *Hum Movt Sci.* 1990;9:241–289.
116. Brady F. A theoretical and empirical review of the contextual interference effect and the learning of motor skills. *Quest.* 1998;50:266–293.
117. Lee TD, Magill RA. Can forgetting facilitate skill acquisition? In: Goodman D, Wilberg RB, Franks IM, eds. *Differing Perspectives on Memory, Learning and Control.* Amsterdam: North-Holland; 1985:3–22.
118. Shea JB, Zimny ST. Context effects in learning movement information. In: Magill RA, ed. *Memory and the Control of Action.* Amsterdam: North-Holland; 1983:345–366.
119. Del Rey P, Whitehurst M, Wood J. Effect of experience and contextual interference on learning and transfer by boys and girls. *Percept Mot Skills.* 1983;56:581–582.
120. Pigott RE, Shapiro DC. Motor schema: the structure of the variability session. *Res Q Exerc Sports.* 1966;55:41–45.
121. French KE, Rink JE, Werner PF. Effects of contextual interference on retention of three volleyball skills. *Percept Mot Skills.* 1990;71:179–186.
122. Farrow D, Maschette W. The effects of contextual interference on children learning forehand tennis groundstrokes. *J Hum Movt Stud.* 1997;33:47–67.
123. Pinto-Zipp G, Gentile AM. Practice schedules in motor learning: children vs. adults. *Soc Neurosci Abstracts.* 1995;21:1620.
124. Shea CH, Kohl R, Indermill C. Contextual interference: contributions of practice. *Acta Psychologica.* 1990;73:145–157.
125. Herbert EP, Landin D, Solmon MA. Practice schedule effects on the performance and learning of low- and high-skilled students: an applied study. *Res Q Exerc Sports.* 1996;67:52–58.
126. Lee E, Gentile AM. Practice schedules in learning a manipulative task: skilled versus novice performers. Manuscript under review.
127. Wulf G, Hob M, Prinz W. Instructions for motor learning: differential effects of internal vs. external focus of attention. *J Mot Behav.* 1998;30:169–179.
128. Maraj B, Allard F, Elliott D. The effect of nonregulatory conditions on triple jump approach run. *Res Q Exerc Sports.* 1998;69:129–135.
129. McCullagh P, Weiss MR, Ross D. Modeling considerations in motor skill acquisition and performance: an integrated approach. In: Pandorf K, ed. *Exercise and Sport Science Reviews, vol. 17.* Baltimore: Williams & Wilkins; 1989:475–513.
130. Shepherd RB, Gentile AM. Standing up: functional relationship between upper body and lower limb segments. *Hum Movt Sci.* 1994;13:817–840.
131. Carr JH, Gentile AM. The effects of arm movement on the biomechanics of standing up. *Hum Movt Sci.* 1994;13:175–193.
132. Schleihauf RE. *The Biomechanical Analysis of Swimming Propulsion in the Sprint Front Crawl Stroke.* New York: Teachers College, Columbia University; 1984. Dissertation.
133. McGuigan FJ, MacCaslin EF. Whole and part methods of learning a perceptual-motor skill. *Am J Psychol.* 1955;68:658–661.
134. Winstein CJ, Gardner ER, McNeal DR, et al. Standing balance training: effect on balance and locomotion in hemiparetic adults. *Arch Phys Med Rehabil.* 1989;70:755–762.

135. Fulton RE. Speed and accuracy in learning movements. *Arch Psychol.* 1945;30:1–53.
136. Solley WH. The effects of verbal instruction of speed and accuracy upon the learning of a motor task. *Res Q Exerc Sports.* 1952;23:231–240.
137. Annett J. Motor imagery: perception or action? *Neuropsychologia.* 1995;33:1395–1417.
138. Schmidt RA. *Motor Learning and Performance – from Principles to Practice.* Champaign, IL: Human Kinetics; 1991.
139. Winstein CJ, Pohl PS, Lewthwaite R. Effects of physical guidance and knowledge of results on motor learning: support for the guidance hypothesis. *Res Q Exerc Sports.* 1994;65:316–323.
140. Winstein CJ, Schmidt RA. Reduced knowledge of results enhances motor skill learning. *J Exp Psychol: Learn Mem Cogn.* 1990;16:677–691.
141. Schmidt RA, Young DE, Swinnen S, et al. Summary knowledge of results for skill acquisition: support for the guidance hypothesis. *J Exp Psychol: Learn Mem Cogn.* 1989;15:352–359.
142. Sidaway B, Moore B, Schoenfelder-Zohdi B. Summary and frequency of KR presentation effects on retention of a motor skill. *Res Q Exerc Sports.* 1991;62:27–32.
143. Salmoni AW, Schmidt RA, Walter CB. Knowledge of results and motor learning: a review and critical appraisal. *Psychol Bull.* 1984;95:355–386.
144. Schmidt RA. Frequent augmented feedback can degrade learning: evidence and interpretations. In: Requin J, Stelmach GE, eds. *Tutorials in Motor Neuroscience.* Dordrecht, The Netherlands: Kluwer; 1991:59–75.
145. Laszlo JI. Motor control and learning: how far do the experimental tasks restrict our theoretical insight? In: Summers JJ, ed. *Approaches to the Study of Motor Control and Learning.* Amsterdam: North-Holland; 1992:47–79.
146. Magill RA, Chamberlin CJ, Hall KG. Verbal knowledge of results as redundant information for learning an anticipation timing task. *Hum Movt Sci.* 1991;10:485–507.
147. Kernodle MW, Carlton LG. Information feedback and the learning of multiple-degree-of-freedom activities. *J Mot Behav.* 1992;24:187–196.
148. Wulf G, Shea CH, Matschiner S. Frequent feedback enhances complex skill learning. *J Mot Behav.* 1998;30:180–192.
149. Janelle CM, Barba DA, Frehlich SG, et al. Maximizing performance feedback effectiveness through videotape replay and a self-controlled learning environment. *Res Q Exerc Sports.* 1997;68:269–279.
150. Swinnen SP, Schmidt RA, Nicholson DE, et al. Information feedback for skill acquisition: instantaneous knowledge of results degrades learning. *J Exp Psychol: Learn Mem Cogn.* 1990;16:706–716.
151. Platz T, Denzler, KB, Mauritz K-H. Motor learning after recovery from hemiparesis. *Neuropsychologia.* 1994;32:1209–1223.
152. Levin MF. Interjoint coordination during pointing movements is disrupted in spastic hemiparesis. *Brain.* 1996;119:281–293.
153. Wolf SL, Lecraw DE, Barton LA, Jann BB. Forced use of hemiparetic upper extremity to reverse the effects of learned nonuse among chronic stroke and head-injured patients. *Exp Neurobiol.* 1989; 104:125–132.
154. Dean CM, Shepherd RB. Task-related training improves performance of seated reaching tasks after stroke. *Stroke.* 1997;28:722–728.
155. Taub E, Wolf SL. Constraint induced movement techniques to facilitate upper extremity use in stroke patients. *Top Stroke Rehabil.* 1997;3:38–61.
156. Crocker MD, MacKay-Lyons M, McDonnell E. Forced use of the upper extremity in cerebral palsy: a single-case design. *Am J Occup Ther.* 1997;51:824–833.
157. Durham P. Distribution of practice as a factor affecting learning and/or performance. *J Mot Behav.* 1976;8:305–307.
158. Ross SL. The effectiveness of mental practice in improving the performance of college trombonists. *J Res Music Educ.* 1985;33:221–230.

159. Ryan ED, Simons J. What is learned in mental practice of motor skills: a test of the cognitive-motor hypothesis. *J Sports Psychol.* 1983;5:419–426.
160. Kohl RM, Ellis SD, Roenker DL. Alternating actual and imagery practice: preliminary theoretical consideration. *Res Q Exerc Sports.* 1992;63:162–170.
161. Driskell JE, Cooper C, Moran A. Does mental practice enhance performance? *J Appl Psychol.* 1994; 79:481–492.
162. Engardt M. Rising and sitting down in stroke patients: auditory feedback and dynamic strength training to enhance symmetrical body weight distribution. *Scand J Rehabil Med.* 1994;Suppl 31:3–57.
163. Moreland J, Thomson MA. Efficacy of EMG biofeedback compared with conventional physical therapy for upper-extremity function for patients following stroke: a research and meta-analysis. *Phys Ther.* 1994;74:534–543.

Chapter 4

Recovery of Function after Brain Damage: Theoretical Implications for Therapeutic Intervention

Jean M. Held

Research on recovery of function has produced some amazing and exciting information that we as therapists will be eager to incorporate into the context within which we practice. Before exploring these research findings, let us return to those explicit and implicit assumptions underlying the neurotherapeutic approaches that were discussed in Chapter 1 and see how these assumptions have shaped our expectations about recovery and therefore about therapy.

Perhaps the assumption that has had the most pervasive influence on our thinking has been that which is based on the hierarchical model of motor control. According to Davis, the principles of the hierarchical model are top-down, unidirectional information flow, localization of function, and executive command.[1] This model leads to the conceptualization of the central nervous system (CNS) as rigid and unable to adapt. Once an area is damaged, its function is lost. Until recently, we thought that regeneration was not possible within the CNS. Therefore, there would be no way for true recovery to occur. If recovery is impossible, then therapy must be aimed at helping a person to compensate for the lost function. For example, we would train a patient with hemiplegia to use the more involved limbs as an assist or prop but to concentrate on the less involved limbs to carry out most daily activities.

Many therapists have never fully accepted these assumptions. We have not set goals of compensation until all hope of return of function of the involved areas has gone. We have seen true return of function. The purpose of this chapter is to explore research findings on recovery of function that can serve to debunk these inadequate assumptions that have until now shaped our therapeutic approaches and that also can serve to guide us in reformulating the basis from which we operate. First, the response of the organism to brain damage will be discussed, both on a neuronal level and a systemic level. This will provide a basis from which to examine the question, "Is regeneration possible within the CNS?" The possible roles of drugs in affecting the responses to damage and/or regeneration will also be reviewed. Second will be a discussion of factors that seem to influence the outcome after brain damage. These factors include certain biological factors such as age and gender of the organism, the nature of the lesion (size, pattern, and momentum), and environment and/or experience. Third, an attempt will be made to extract implications for therapeutic intervention.

It is necessary to define certain terms before continuing. *Function* can be defined in two ways: the activity performed by a given organ or the whole organism's complex activity directed at performing some behavioral task. The latter definition is more appropriate in studying recovery of function. However, different investigators have failed to discriminate between function as simply goal attainment and function that is characterized by the pattern of movement and its efficiency in accomplishing the goal. Only by making this distinction can we hope to recognize the difference between true restitution of function and merely behavioral substitution. *Recovery* is the gradual return of a specific function after an initial deficit is observed following CNS damage. To be considered truly recovered, the function must be performed in the same manner and with the same efficiency and effectiveness as before the damage. *Sparing,* another term frequently used when discussing recovery of function, is the absence of deficit immediately following CNS damage or less deficit than is customarily observed.

RESPONSES TO CENTRAL NERVOUS SYSTEM DAMAGE

Neuronal Responses

Any physical injury that causes neurons in the adult CNS to die will cause a permanent change in the structure of the CNS because neurons have withdrawn from the mitotic cycle and are no longer capable of cell division.[2] However, many injuries involve interruption of axons rather than direct trauma to cell bodies (Figure 4–1). If an axon is severed, the mechanisms that carry materials synthesized in the cell body to the axon terminals are interrupted. As a

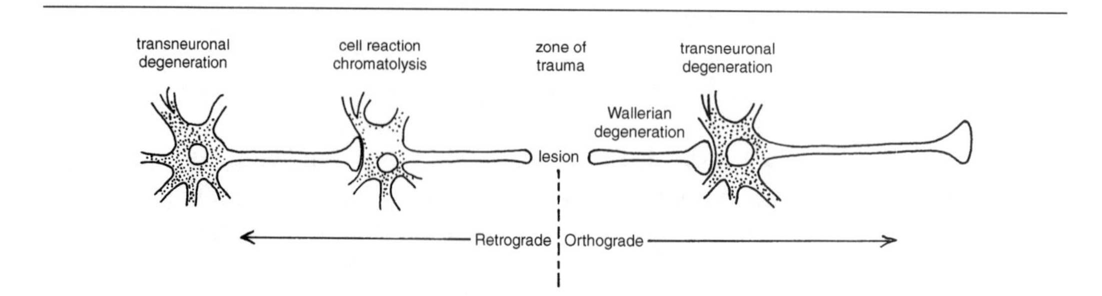

Figure 4–1 Neuronal response to damage.

result, the axon and terminals are deprived of their normal metabolic connection with the cell body and they degenerate. When the axon is cut, the ends of the proximal and distal segment seal off the axoplasm, retract from one another, and swell. In the zone of trauma, there is rapid degeneration of the axon and the myelin sheath. Given that blood vessels are usually interrupted by the lesion, thereby breaking down the blood-brain barrier, circulating macrophages absorb and destroy axonal debris. Glial cells also proliferate and act as phagocytes. The proliferation of fibrous astrocytes leads to the formation of glial scarring around the zone of trauma. Degeneration spreads in both directions along the axon from the zone of trauma. Proximally, this usually proceeds to the point of origin of the first axon collateral.[2]

In addition to degeneration of the injured axon, there may also be transneuronal degeneration, either orthograde or retrograde. This secondary neuronal death can occur within 2 to 7 days of the injury and can account for more damage than primary cell death. With orthograde degeneration, the degree of atrophy of a pathway is related to the percent of total input removed from the population of neurons by the lesion. It is thought that other sources of input can reduce the severity of transneuronal degeneration. Proposed mechanisms to explain orthograde transneuronal degeneration are that the nerve requires a certain amount of stimulation to survive or synaptic terminals normally release trophic substances that are necessary for survival. Retrograde transneuronal degeneration has also been observed; however, this is more difficult to explain. Both types of transneuronal degeneration may occur over more than one synapse and thus may have far-reaching effects.[3]

One illness that often results in neuronal cell death is stroke. Stroke may be caused by hemorrhage into the brain (20% of cases) or ischemia (80% of cases). With ischemia, the amount of damage is related to the length of time that ischemia persists.[4] Neurons experiencing severe ischemia die within 5 minutes. Surrounding neurons may die from a series of reactions set off by the initial ischemia. Following the lack of oxygen and nutrients, neuronal membranes are depolarized, which leads to the release of glutamate and other amino acids.[5] The effect of glutamate on the postsynaptic receptors increases intracellular calcium and produces oxygen free radicals. This combination leads to rapid cell death.

Systemic Responses

The Russian neurologist Alexander Luria and his associates suggest that focal brain lesions produce two types of functional disturbances: (1) those resulting from actual death of neurons, whose functions disappear completely, and (2) those resulting from inhibition of intact neurons.[6] This latter concept of inhibition of intact neurons is not a new one. Von Monakow coined the term *diaschisis* to explain transient deficits following brain damage.[7] He thought that diaschisis was caused by a functional standstill or abolition of excitability transmitted to neural areas related to the damaged part of the system. He thought that inhibition of function would disappear over time. This is reminiscent of current concepts of neural shock. Goldberger and Murray associate the idea of diaschisis with the decreased excitatory state of postsynaptic neurons following partial denervation.[8]

Luria and colleagues state that the relative proportions of these two components (cell death and inhibition) will depend on the type of injury.[6] For instance, a massive intracerebral hemorrhage will cause widespread destruction of nerve cells as well as more distant inhibition (or a decreased excitatory state), whereas a concussion may cause a much higher proportion of inhibition than cell death. Treatment must consider this information. When a decreased excitatory state is the key, therapy must disinhibit the system.

Le Vere views the systemic response to damage a little more generally.[9] He thinks that a person's behavior subsequent to brain injury is an attempt to respond on the basis of the best neural system currently available. This logically precludes those neural systems directly affected by the injury, regardless of how much of the system is spared. The patient will therefore attempt to compensate for behaviors mediated by the damaged system. If compensation leads to goal attainment, then there is no need for recovery of the damaged system because the needs are met. If compensation does not work and there is enough of the damaged neural system left intact, recovery will occur. LeVere thinks, therefore, that compensation interferes with postlesion expression of spared neural mechanisms underlying recovery. In other words, if compensation is allowed to occur, then there is apparently no stimulus to the partially damaged system to recover and behavioral substitution will occur.

Taub suggests that a critical period may come immediately after CNS damage occurs when, if the patient attempts to move the involved limbs, he or she will be unsuccessful.[10] This may be from the diaschisis or neural shock. This failure to be successful may lead the patient to develop compensatory strategies and to develop what Taub refers to as learned nonuse. However, forced use during this period may be detrimental. Schallert and colleagues found that forcing rats with cortical lesions to use the involved limb immediately after surgery led to larger lesions (a greater amount of cell death) and permanent deficits.[11] But, if the patient is allowed to continue to use compensatory strategies after the period of diaschisis has subsided, the patient will not recover. Nudo and colleagues found that with training of the involved limb of monkeys following cortical lesions, the motor representation formerly present in the lesioned area spread to adjacent regions, and behavioral recovery occurred.[12] However, in Nudo's studies, the training was delayed by several days after the initial lesion. Thus, the neuronal and systemic response to damage will depend on the demand placed on the system both during the initial period of diaschisis or neural shock and during the period following the resolution of diaschisis.

REGENERATION IN THE CENTRAL NERVOUS SYSTEM

As early as the turn of the century it was suggested that some axonal regrowth was possible within the CNS, but, because it was limited, it would therefore be of questionable functional significance. Even as recently as the 1960s, research demonstrated that although regeneration occurred the new sprouts failed to cross the damaged area and make functional connections. It is known now that two types of sprouting or synaptogenesis occur: (1) collateral sprouting or reactive synaptogenesis and (2) regenerative sprouting or synaptogenesis (Figure 4–2). Numerous other factors affect the success of either of these types of sprouting. In addition, some investigators have suggested that there may be a way to replace lost neurons, eg, by transplantation. Let us explore these three mechanisms.

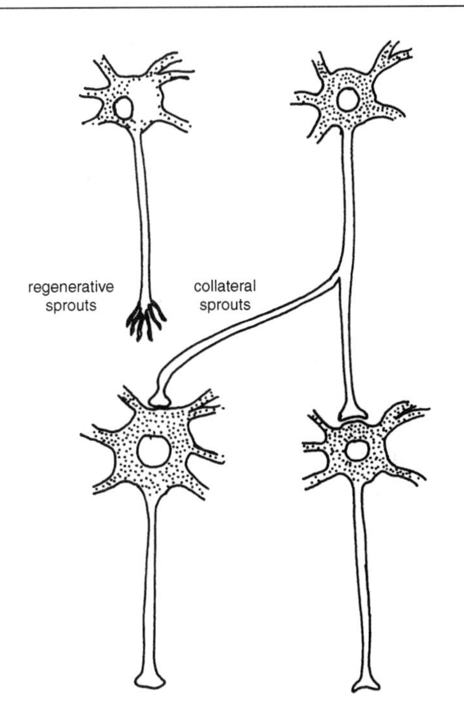

Figure 4–2 Synaptogenesis.

Reactive Synaptogenesis

Within 4 to 5 days of injury, sprouts from nearby undamaged axons appear and repopulate nearby vacant fields. This repopulation does not replace the original circuitry but results in an increase in residual inputs. This reactive synaptogenesis is not random. Most of the new inputs are from systems most closely related to the injured area. This may be beneficial in two ways: (1) it may prevent dendritic atrophy resulting from denervation, and (2) it may maintain a functional level of excitability so that other inputs may be effective in activating the pathway. However, reactive synaptogenesis may create abnormal connections or interfere with recovery by competing with regenerating sprouts for their normal synaptic connections. Nonetheless, reactive synaptogenesis proves that new synapses can be formed in the mature CNS, replace those lost by injury, and mediate synaptic transmission.[3]

Regenerative Synaptogenesis

About the same time that reactive synaptogenesis occurs, regenerative sprouts start to grow from cut axons near the injury site. These sprouts must travel over a long distance, often longer than collateral sprouts, and usually do not succeed in reconnecting to their original targets. Although it was once thought that glial scarring prevented the regenerative sprouts from reaching their destinations, it is now thought that the problem is more complex. The glial cells in the area may not produce enough neuronotrophic substances, or they may produce substances that would inhibit growth. A great deal of research is being conducted to find ways to improve the environment to enhance the possibility of the regenerative sprouts reinnervating their usual targets.

NEURONAL REPLACEMENT VIA TRANSPLANTATION

Neurogenesis does not appear to occur in the adult mammalian CNS. Transplantation of fetal CNS has been proposed as an alternative way to replace neurons lost by injury. It has been a useful model for studying the requirements for successful neuronal survival. Embryonic brain tissue transplants have been used both to stimulate regeneration as well as to replace lost cells. These transplants have the remarkable capacity to grow long distances and connect with appropriate targets.[13] The ability of these transplants to aid in function would appear to be related to whether the reinnervation pattern and its pharmacology match the original properties of the system under study.[13] This neurotransmitter identity seems to be one of the main specificity determinants for afferent fibers. Such transplants appear to act as an endogenous source of additional transmitter and perhaps trophic factors that are delivered in sufficient quantity to appropriate receptors. They do not, however, fully reform original circuitry. They appear to act primarily by delivering chemicals to the proper area much as an endogenous drug delivery system.[3]

Transplants have been found to bring about profound behavioral restoration. In an animal model of Parkinson's disease, several different types of grafts have led to behavioral improvements of recovery.[14] By injecting a neurotoxin, 6-hydroxydopamine (6-OHDA), relatively selective and permanent lesions of dopamine neurons can be made. Bilateral lesions placed in the nigrostriatal pathway in rats produce akinesia, hunched posture, poverty of initiation of voluntary movement, and severe aphagia and adipsia. Unilateral lesions induce an asymmetrical postural bias toward the side of the lesion and a hemiparkinsonian syndrome. With grafts of pieces of embryonic substantia nigra into the lateral ventricle, 50% compensation for motor abnormalities was seen. When the transplant consisted of a piece of embryonic mesencephalon placed in a cortical cavity in direct contact with the dorsal or lateral surface of the neostriatum, partial to complete compensation was observed. The degree of compensation was directly correlated with the extent of dopamine fiber ingrowth from the transplant. When suspension of fetal nigral cells is injected into the neostriatum via stereotaxic apparatus, 100% recovery of behavioral impairments developed. Finally, because of the potential problems with obtaining embryonic tissue, alternative peripheral nervous tissue was tried. Adrenal medullary cells were placed into the lateral ventricle, and 40% to 50% compensation occurred.[14] Following up on the animal studies, the latter procedure was performed on human patients with Parkinson's disease for the first time in 1985.[15] In the first patient, who suffered from severe, progressive symptoms, no change occurred in clinical status. However, this constituted a marked difference from the previous rapid decline. In the second patient twice as much tissue was implanted. This patient was more severely impaired preoperatively than the first patient. Following the surgery she exhibited a decreased rigidity in the upper extremities, an increase in voluntary movements, and more normal facial expression. After 4 months, the person's clinical status stabilized but did not worsen. This suggests that the effects of the implants wear off with time.[15] Since those initial transplants of adrenal cortical tissues in humans with Parkinson's disease, procedures have been developed to transplant human fetal substantia nigra tissue. The number of patients who have received neural transplantation is not known. However, most cases report improvements, although none had complete remission.[16,17]

Fetal brain tissue transplants have been carried out in other regions of the brain in animals with behavioral consequences. In a series of studies, Stein and colleagues studied the effects of fetal transplants in the frontal cortex on the acquisition of a spatial alternation task after prior lesions of the frontal cortex.[18] In addition, they studied the effects of transplants in the visual cortex on the learning of a brightness discrimination task after prior lesions to the visual cortex. In both cases, rats with the transplants performed the respective tasks better than animals that sustained the same cortical lesions, although not as well as the sham-operated controls. In the visual cortex study, embryonic frontal cortex transplants were more effective than visual cortex transplants in ameliorating deficits in brightness discrimination. Stein and colleagues thus concluded that the specificity of fetal tissue is not critical for promoting function recovery in brain-damaged rats.[18] Instead, the immaturity of the fetal tissue may be more critical. Dunnett and colleagues reported that animals who received fetal transplants following cortical damage performed worse than animals who had lesions alone.[19] This could be from the transplants producing larger lesions, and thus poorer performance over time, or to transplants interfering with the normal recovery processes.

Numerous issues surround the use of fetal tissue transplantation: some ethical and some procedural. Some animal studies have shown that the transplants do not survive, and lead to larger lesions. Others show that transplants overgrow and cause further damage by becoming a space occupying growth, causing increased intracranial pressure. There is some suggestion that there may be an immune reaction to the transplants, thus sealing off the transplant, and preventing connections from being formed with the host brain. Some studies have demonstrated no benefit to the use of fetal tissue transplants. Thus, the use of transplants in humans must be advanced carefully.

PHARMACOLOGICAL INTERVENTION

Pharmaceutical agents may be used in several different ways to try to affect recovery after brain damage. On the cellular level, drugs may be effective in preventing the neurotoxic events following damage, preventing glial scarring and stimulating growth. On the systemic level, drugs may be used to affect the overall level of excitability of the systems that have been partially denervated.

Prevention of Neurotoxicity Following Damage

As mentioned earlier, numerous biochemical events that follow initial CNS injuries can be detrimental to the patient. Numerous studies have been done to find agents to interrupt this sequence of events and thus prevent the secondary neuronal cell death that results. Basically three types of drugs are being investigated: (1) neuroprotective agents, (2) neuroreparative agents, and (3) anticoagulant and defibrinogenating agents.[20] The neuroprotective agents are intended to do one of the following: block calcium channels to prevent the buildup of intracellular calcium; prevent the excessive extracellular glutamate; prevent the damage from oxygen free radicals; or inhibit glutamate. Neuroreparative agents promote the repair of neurons. Anticoagulants and defibrinogenating agents promote the reperfusion of the ischemic areas. Some animal research has been encouraging; however, in studies in humans following stroke, the results have been equivocal. First, diagnosis confirming that the stroke is the ischemic type is critical, as many drugs under study can cause hemorrhage and are extremely detrimental in patients whose stroke is of the hemorrhagic type. Second, treatment must occur within 3 to 4 hours of the injury, and it is only effective if the area that was ischemic is reperfused. Third, 25% of patients with stroke have extensive damage and will likely have severe permanent disability. Drug intervention in these patients will likely not improve their outcome and may prolong their life. Fourth, the drugs are extremely expensive.[4] Thus, additional research is necessary before standard protocols using these types of pharmacologic agents can be developed.

Prevention of Scarring

Given that glial scarring was once thought to be the primary limiting factor preventing successful regeneration, it was logical that an attempt be made to find drugs that would decrease or eliminate scar formation. In the 1950s Windel and colleagues used a pyrogenic substance derived from *Pseudomonas* to treat spinal cord-transected cats and examined the extent of scarring and growth of axons across the area of damage. Within the first month a clear difference was seen in the amount of scarring between treated and untreated cats.[21] In the treated cats there was more vascularization and less collagenous connective tissue than in the untreated group. Sprouting appeared at about the same time in both groups, but in the untreated group the regenerating neurons were deflected by the glial scars, whereas in the treated group sprouts grew through the lesion site.[21] However, in this study no attempt was made to evaluate behavioral recovery. In later attempts to replicate these findings, a few studies reported complete return of motor and sensory function but many others showed no return of function.

Stimulation of Growth

Given that the reduction of glial scarring has been relatively unsuccessful in producing behavioral recovery and because it is no longer thought that glial scarring is the prohibitive factor to successful regeneration, attention has shifted to an attempt to stimulate growth. One area of study has been the effects of hormones on regrowth. Harvey and Strebnik investigated the effects of thyroxine on the regeneration of spinal nerves following compression injury to the cord in rats.[21] Injections were given to the animals either before or after injury, and treated rats in both instances did better than untreated rats, although never as well as normal animals. On histological examination of the cord in the thyroxine-treated rats, scar tissue was thin or absent and fibers from the ventral or lateral columns grew across the crushed segments. The investigators speculated that thyroxine increased protein synthesis, decreased scarring, and increased vascularization.[21]

In the Soviet Union, researchers used enzymes such as elastase or hyaluronidase to break up scar tissue after spinal cord injuries in rats and have claimed remarkable success.[21] About 44% of the animals reportedly showed complete functional recovery, with looser scar formation and better vascularization near the injured area. However, later attempts to replicate these findings have—for the most part—failed.

Many neurotrophic substances have been discovered and studied as to their effects on neuronal survival and/or regeneration. Nerve growth factor (NGF) was discovered accidentally in 1948 and has since been studied for its ability to enhance recovery after damage to different brain sites.[22] Among many others, Hart, Chaimas, Moore, and Stein[23] studied the role of NGF in facilitating recovery, in this case of learned habits after lesions of the caudate-putamen complex. After bilateral, simultaneous damage, one group of rats received bilateral injections of NGF. Initially after surgery, both groups of brain-damaged animals were functionally impaired. With training, the rats with one dose of NGF at surgery performed at the level of sham-operated controls, and the untreated group continued to show a learning deficit for as long as they were tested. Such behavioral recovery after brain damage and NGF treatment has also been demonstrated in other brain regions. In the study of Hart and colleagues, histological examination did not reveal neuronal sparing; however, there was greater glia-to-neuron ratio for untreated rats as compared with rats treated with NGF.[23] This suggests that there may have been less glial scarring.

Gangliosides are another category of neuronotrophic substances that are found in high concentrations in the CNS. Gangliosides promote sprouting, enhance the effects of NGF, reduce cerebral edema, and protect the sodium/potassium pump.[24] Gangliosides have also been shown to reduce initial deficits after brain injury, speed recovery, and reduce mortality in animals as well as to improve function in humans.[25–27]

Neurotrophic factors have been found to exist in the normal CNS, with an increase in concentration as a response to injury to the CNS.[3] A clearly time-dependent increase in neurotrophic factors occurs in the developing, mature, and aged rat brain and can be elicited by mechanical, chemical, or ischemic damage. Maximal levels of trophic activity are progressively higher in neonatal, adult, and aged animals and are reached progressively later in the respective groups. Such neurotrophic factors have been found after damage to several different brain sites. By using fetal tissue transplants to study the effects of these trophic factors, it was found that if transplantation is timed to coincide with maximal production of neurotrophic factors, then survival and integration with the host are greatly enhanced.[3]

Disinhibiting the System

Very early reports from the 1940s showed that certain stimulant drugs could enhance the rate and degree of recovery from motor cortex damage in monkeys.[28] This supports notions of von Monakow and Luria that part of the results of damage is a decrease in excitability in related systems. In the 1960s Meyer and colleagues removed the entire neocortex of adult cats and tested the animals for placing responses for 1 year after the lesions, with no responses observed.[29] They then administered amphetamine, and within 10 to 20 minutes the cats demonstrated placing responses once again. When the drug wore off; the deficits returned. They suggested that lesions cause a suppression of the retrieval of "engrams" that have been formed in the brain as a result of learning or experience. Thus, memory traces are not destroyed by cortical lesions but access to them may be blocked. Amphetamine somehow allows the animals to overcome the block.[29]

The results of Stein's experiments may be seen as further support for Meyer's theory.[21] Stein unilaterally ablated the posterior parietal cortex in mature Java monkeys and found that the monkeys were unable to use the affected hand to reach for a small piece of banana and place it in the mouth. After 10 days, recovery was evident. He then created a second lesion in the contralateral cortex. Again, recovery was observed within 10 days to 2 weeks. One year later, these same monkeys were given a small dose of anesthesia. An immediate return of the deficit was seen. As the drug wore off, normal behavior returned. Stein hypothesized that with training, the level of arousal was sufficient to permit "access to the engram." When arousal was diminished, access was again blocked. Feeney and Sutton demonstrated that the administration of a single dosage of d-amphetamine following sensorimotor cortex removals in rats and cats resulted in an immediate recovery of constrained locomotion, given testing/training during the period of drug intoxication.[24] However, a single dose of haloperidol, a depressant, impeded recovery of constrained locomotion for several weeks.

Meyer and Meyer hypothesize that the substrate for good performance in brain-damaged subjects may lie dormant.[30] Thus, therapy should involve a combination of treatments with pharmacological agents and behavioral procedures for the evocation of the memories to be reinstated. Therefore, therapy would not help patients to recover by forming new habits but instead to make better use of habits they already have.

Few attempts have been made to manipulate recovery with drugs in humans owing to the potential dangers of stimulants in patients with CNS damage.

However, Luria and associates administered anticholinesterase drugs to brain-damaged subjects and reported that such treatments had remarkable, dramatic effects, often restoring lost functions immediately, even when disease had lasted for years.[6] They reported that improvement was stable and served as the foundation for later exercise therapy. More recently, the effects of d-amphetamine were studied in patients with stroke. Persons received a single dose of d-amphetamine or a placebo, followed immediately by physical therapy. Those persons that had received the d-amphetamine performed better than the placebo group.[31]

FACTORS AFFECTING SPARING AND RECOVERY OF FUNCTION

Many factors have been found to affect the results of damage to the CNS as well as the recovery seen thereafter. These factors may be divided into biological factors, characteristics of the lesion, and experiential factors such as environmental exposures or training.

Biological Factors

Age

Until recently, a rather universal opinion, sometimes called the Kennard principle, has greatly influenced both researchers of recovery of function as well as clinicians. This principle holds that brain damage sustained during infancy causes fewer deficits than comparable damage in adulthood. However, out of recent studies and a reevaluation of classical evidence, a more complex pattern emerges. First, damage to certain areas during infancy produces the same deficits as in adulthood. Second, although considerable sparing may be evident initially, deficits often develop as the organism matures.

It has been suggested that the critical variables determining deficit versus sparing are the maturational status of the area damaged, the functional status of the remaining systems, the size of the lesion (or relative percentage of the system), and the experience of the organism before the damage.[32] If an area is functionally mature, damage will result in comparable deficits with those seen in adults. However, if a functionally related area is not mature in that it is not yet committed to its own function, it might take over the function of the damaged tissue during development. On the other hand, if the area damaged had *not* reached functional maturity at the time of damage, no deficit would be apparent at first but would develop at the time when the area would normally assume its function during maturation.

Sparing or recovery of function brought about by other areas taking over may be at some cost in that there may be some crowding of the remaining intact structures as they attempt to subserve more functions than normally required. For example, language functions may be spared in young children with left hemisphere damage, but this sparing is often accompanied by diminished capacity for spatial perception, which is thought to be a function of the right hemisphere.[33]

Also, brain damage early in life may result in greater plasticity, causing anomalous growth in the form of synapses and/or tracts.[34] These may approximate normal control of function if they reproduce normal circuitry; however, this anomalous growth may lead to abnormal functions as well. Thus young children who sustained prenatal or perinatal brain injury appear to demonstrate less spasticity than adults who sustain the same type of damage; however, spasticity and its often concomitant paresis become obvious later in development (4 to 12 months).[35,36]

The same consideration is appropriate in the case when brain damage occurs in the aged organism. Areas may serve different functions in the aged system than in the newborn or the adult; thus, deficits will not be the same with lesions occurring at different ages. A study supporting these notions was conducted by Stein and Firl.[37] Four groups of animals (two young, sexually mature male rats and two aged male rats) were tested on a delayed spatial alternation task. One group of young rats and one group of aged rats received bilateral frontal cortex ablations and were compared postoperatively with each other and with the age-matched, nonlesioned groups. Despite the lesions in one group, both groups of aged rats learned the task in about the same time but were slower than the young rats that were not operated on. The lesioned young rats, however, were very impaired on the task. On histological examination, it was shown that the young nonlesioned rats had six times as many cells in the dorsomedial nucleus of thalamus, the origin of projections to frontal cortex, as the aged nonlesioned rats. However, there were no differences in the number of cells in the lesioned versus nonlesioned aged rats. These results led Stein and Firl to speculate that the frontal cortex may serve different functions throughout the lifespan.[37]

Gender

It is becoming clearer with time that a difference exists in the organization of the CNS depending on the sex of the organism. For example, it has been demonstrated that females are less strongly lateralized in cortical functions than are males. If this is the case there may also be differences in the response to and potential recovery after brain damage. Researchers are just beginning to explore this hypothesis. Caulder and Gentile demonstrated, in a study on recovery of constrained locomotion following sensorimotor cortex damage in rats, that females performed significantly better than males. In addition, males who were gonadectomized before the brain damage performed as well as the females.[38] However, Roof and colleagues studied the effect of hormones on the formation of edema following brain damage. They found in a series of studies that the presence of progesterone led to lesser amounts of edema.[39] On the clinical level, little evidence shows that either gender recovers better following brain damage.

Characteristics of the Lesion

Three different aspects of the characteristics of lesions appear to influence the effects of damage and the potential recovery. These considerations are the size of the lesion, the apportionment of symmetry of the damage, and the momentum or rapidity with which the damage was sustained.

Size

In general, the smaller the lesion, the less the deficit and the greater the recovery. However, the smallness or largeness depends on the area of the brain involved. The critical considerations are whether a functional area is entirely or partially removed and how tightly coupled that area is. A great deal of evidence indicates that when even a tiny proportion of a functional region is left intact, recovery is possible. Also, certain systems are more diffusely organized. An example of the latter type of system is the dopaminergic system of the basal ganglia and the nigrostriatal pathway. The fact that symptoms can be temporarily ameliorated by L-dopa suggests that the dopamine system exerts a primarily permissive function.[23] It enables other striatal neuronal processing to occur and regulates the general level of striatal activity.

Momentum and Apportionment

It is generally the case that slowly developing lesions in the CNS create less disruption of function than lesions of the same site and extent that are sustained suddenly. Numerous case studies have been cited throughout the years in which a person was functioning well until very near his or her death, after which the autopsy revealed a huge area of completely necrotic brain tissue.

In the animal research realm, in an attempt to develop a model to study this phenomenon, serial lesions are created under controlled conditions (Figure 4–3). It seems that a number of conditions affect what has become known as the *serial lesion phenomenon.* First, the amount of tissue removed in each stage is important. The less tissue removed at one time, the less disruption to the system despite the possible complications of multiple surgeries, anesthetizations, and the fact that serially placed damage usually exceeds the size of one-stage damage.[40] Second, the interoperative interval is critical. There seems to be a minimal separation necessary to differentiate between one-stage and two-stage lesions.[41] Third, interoperative experience is significant. Postoperative care certainly

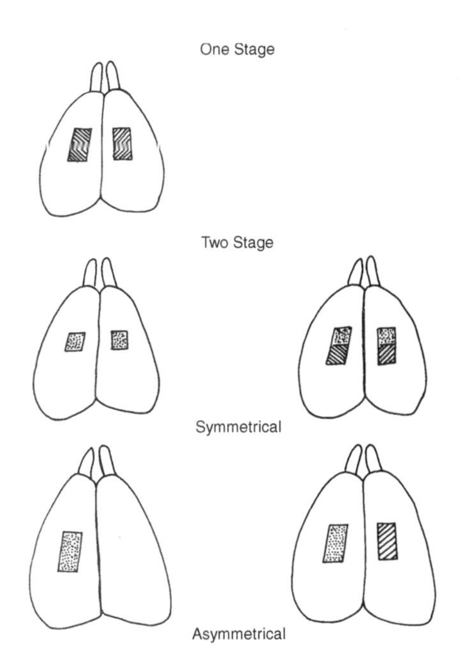

Figure 4–3 Serial lesion paradigm. Stippled areas are removed during first operation and striped areas are removed during second operation.

influences the outcome.[42] In addition, the general level of stimulation, unrelated to the specific behavior under study, seems to be of note. The presence of light versus total darkness and the presence of sound versus silence have been shown to affect whether animals sustaining serial lesions perform any better than one-stage animals.[40] Further, in certain cases, interoperative training is necessary to demonstrate the serial lesion effect.[41] Fourth, the pattern of placement of the lesions influences the outcome. For the most part, symmetrically placed lesions produce less disruption than asymmetrically placed lesions (assuming the amount of tissue removed is comparable).[42]

Several explanations have been offered for the serial lesion phenomenon. It has been suggested that diaschisis is greater when the total damage occurs suddenly or all at once. In addition, the slowness may lead to optimal conditions for sprouting or denervation supersensitivity. Interestingly, the minimum interoperative period to produce the phenomenon (10 to 14 days) coincides with the time period for sprouting to occur (12 to 15 days). Also, it has been suggested that the gradual onset leaves time to learn to use different cues or strategies to accomplish the task.[40]

Experiential Factors

The idea that experience can influence an organism's response to brain damage relates to the controversy as to the importance of nature (or genetically determined, hardwired mechanisms) versus nurture. In 1949 Hebb posed the question whether performance could be improved by providing laboratory rats with more than a routine existence in semi-isolation in small metal cages. To address this question, he took two litters of rats home and raised them as pets. He later tested both the "pets" and the rats raised in standard laboratory conditions on maze-learning tasks and found that all of the "pets" scored in the top one-third of the total distribution and continued to improve their relative standing. Hebb interpreted these results by claiming that the richer experience during development made them better able to profit by new experiences at maturity—one of the characteristics of the intelligent human being.[43] Stated in another way, these animals had improved problem-solving abilities.

However, impoverished conditions or deprivation can mimic actual brain damage. Children raised in foundling homes were studied in terms of the effect on later capabilities. A small group of children were removed from the home and placed with girls with mental retardation in a setting where they received physical care, love, attention, and stimulation. When the children were removed from the foundling home, their IQs were in the 60s. On follow-up it was discovered that their IQs had risen to normal in the new environment, and later the grown children were found to be leading normal, productive lives. Those children not removed from the foundling home at an early age had to be cared for in one type of institution or another as adults.[21] Similar evidence exists in controlled animal studies. When newborn kittens are deprived of certain types of visual experience, such as when one eyelid is sutured closed, they fail to develop normal perceptual capacities, in this case depth perception.[44]

Environment

Studies on the exposure of animals, particularly rats, to enriched environments revealed that many morphological and biochemical changes result: increased cortical depth, increased brain weight, increased dendritic branching, and increased enzyme activity.[45] Given these findings, researchers began to study whether this enrichment could affect responses to brain damage. Two possibilities exist: (1) enrichment before brain damage and (2) enrichment after brain damage.

In animal studies, preoperative environmental enrichment has been found to have a protective effect against deficits after brain damage. In rats who received lesions of anterior or posterior cortex, those who were exposed to enrichment before surgery made fewer errors on the maze learning problems than those who were isolated in individual cages. Enriched animals with cortical lesions performed better than restricted animals without brain damage.[46] This same benefit of preinjury enrichment has been demonstrated with hippocampal, septal, and sensorimotor cortex damage, although not in all cases was the protection complete.[47–49]

Postoperative effects of enriched environments have been studied more extensively. Again, the enrichment led to either less deficit or greater recovery than impoverishment after brain damage in several regions of the brain, and in some cases it was demonstrated that the enriched rats were not only better than the impoverished, lesioned rats, but better than impoverished sham-lesioned rats.[49–52]

Held, Gordon, and Gentile examined the effects of preoperative and/or postoperative enrichment versus impoverishment on constrained locomotion following sensorimotor cortex removals. Animals that were preoperatively enriched were not only indistinguishable from enriched, sham-lesioned controls on the basis of behavioral measures, but also fine-grained movement analysis revealed that their movement patterns were normal.[49] Those animals that were postoperatively enriched without preoperative exposure were mildly impaired on the locomotor task, although they recovered to preoperative performance quicker than impoverished, lesioned animals. However, after full recovery on the basis of the behavioral measure, their movement patterns were shown to be aberrant. The results of this study suggest that preoperative environmental enrichment has powerful capabilities to protect the organism from the usual results of brain injury. Although postoperative enrichment is also effective, it does not yield the same degree of recovery, at least in terms of the specific motor patterns used.[49]

Finger has suggested that nonspecific environmental enrichment may affect a general factor or a host of specific factors that may relate to the "general adaptive capacity" of the organism, ie, its ability to respond appropriately and to cope with a variety of situations and problems of a general nature.[53] Other researchers think that the structural and biochemical changes that follow enrichment play a role in recovery.[51] Enriched subjects have developed functional neural circuitry that the restricted subjects do not have, and this may provide them with a greater capacity for reorganization after brain lesions; or it may allow these subjects to use existing connections, established independently of the injury, to perform better on certain kinds of tasks.

Clinically, the setting in which a patient receives care and treatment after brain damage would be the equivalent to the environment in the animal studies. Few studies have attempted to evaluate the effects of different settings on recovery. One study carried out in England compared stroke patients that were treated in general wards versus those treated in stroke rehabilitation units.[54]

Despite less physical therapy and equivalent occupational therapy on the stroke units, those patients recovered sooner and regained greater function than those treated on the wards.

Training

Training, or therapy, differs from enriched environmental exposure in that the activities are specific rather than nonspecific or general. Meager evidence indicates that specific training before brain injury may lead to sparing of the function after injury. In animal research, overtraining on a specific task can lead to nearly total sparing of these skills after brain damage.[55] The same is true when conditioning is used to train a task preoperatively.[56] However, in both instances, the sparing is specific to the task learned and does not seem to transfer to similar tasks during the animals' spontaneous movements. Nudo and colleagues studied the effects of training on motor representations following sensorimotor cortex damage in the regions representing the forelimb digits.[12] They found that without training, no evidence indicated representation spread to surrounding areas; however, with training, the representation formerly in the lesioned areas spread into the surrounding tissues, and recovery of function of the hand occurred. Thus, it is thought that with training, the reorganization of the cortical representations can be influenced.

Few clinical studies have been done on the effectiveness of therapy for the patient with neurological impairments, and most of these have been negative. However, one of the earliest animal studies was done by Ogden and Franz.[57] They destroyed motor cortex to produce hemiplegia in monkeys. They then varied postoperative management as follows: (a) no treatment; (b) general massage to the involved limb; (c) restraint of the noninvolved arm; and (d) restraint of the noninvolved arm, massage, and stimulation of the involved limbs to move (using largely noxious stimuli at first), specific facilitation to more involved muscle groups, and forced active movement of the animal. Only in the last condition did the monkeys show recovery, and the recovery was—by the researchers' description—complete within approximately 3 weeks. Black and colleagues examined the effects of specific motor training in monkeys after the cortical precentral forelimb area was surgically removed on one side.[58] Training was varied as follows: (a) contralateral forelimb, (b) ipsilateral forelimb, or (c) both forelimbs, initiated immediately or after 4 months and lasting 6 months. Training of the weak hand alone or the weak and the normal hands together resulted in better recovery in the weak hand than did training in the normal hand alone. Training of both hands was no better than training of the weak hand only. When training was delayed, final recovery was poorer than for those receiving immediate training. Several other investigators have documented the need postoperatively for passive movements and functioning, forced use of a limb, and specific training.

As mentioned in an earlier section, training begun too early may be detrimental. In Schallert's work, forced use of the impaired limb immediately following brain injury in rats led to larger lesions and a permanent deficit.[11] However, Nudo's studies demonstrated that specific training following cortical lesions in the hand region led to behavioral recovery; their training was delayed by 5 days.[12] Thus a critical period may exist when vigorous training should be avoided.

From the sparse evidence on training before damage, it appears that the state of the system at the time of damage is important in determining the response to damage.[59,60] One might suggest that it relates to learning to learn, which is still an enigma to researchers and theoreticians alike, or to actual structural or biochemical changes in neural tissue affecting broad areas of the system and providing the basis for more efficient adaptation to localized injury.

Training or therapy after damage has also been shown to improve recovery: early training after injury is better, as long as it is not too early; and the more specific the training, in some instances, the better. One would expect that the effects of postinjury training may be related to ensuring a functional level of excitability within the system and perhaps preventing transneuronal degeneration. In addition, specific training may prevent the use of compensation, thus stimulating the remaining intact part of the system to recover and avoiding learned nonuse.

IMPLICATIONS FOR THERAPEUTIC INTERVENTION

The fact that sparing and recovery have been observed immediately after brain damage brings into question the hierarchical model of motor control, with its principle of localization of function. Lashley's notions of equipotentiality may be a little extreme because it does appear that certain parts of the CNS are more important to specific aspects of behavior than others.[61] However, clear evidence indicates that CNS function is more complex, with a cooperative style of control rather than executive command.[21] In addition, regeneration can and does occur in the CNS. However, it would appear that for the regeneration to be successful in subserving normal function, the system needs to be helped along. Intervention is needed to prevent secondary neuronal death, control sprouting and guide the growth of regenerating sprouts, aid in axons successfully reaching their targets, and perhaps to replace lost neurons.[3] What are the implications for physical therapy?

First, we as therapists need to change our operating assumptions. We should *expect* recovery and work for that by preventing rather than encouraging compensation to occur. We should carefully analyze each person's problems, and we must intervene early.

Second, we should examine carefully the environment we create for patients. Is it enriched, ie, stimulating, varied, interesting, challenging, and unique? Or is it impoverished, ie, quiet, all the same, undemanding, and depressing? Evidence suggests that environmental enrichment is effective in enhancing recovery after brain damage in all age groups. We should not, therefore, confine our thoughts of enrichment to children but extend them to adults, including the aging adults. Take a careful look at your hospital. What are the rooms like where patients spend most of their time? Is one room different from the next? Is there anything uniquely personal for each patient? What about the therapy departments? Are there bright colors on the walls, pictures, interesting and functionally related tasks to do, and interesting people with whom to interact (do all therapists look alike)? We should become instrumental in ensuring that the environment is at least as interesting as the average person's environment, and even better, that the environment is especially interesting and involving. The environment may help to disinhibit the suppressed system; prevent transneuronal degeneration; and stimulate sprouting, dendritic branching, and increased enzyme activity.

Third, we need to provide early, specific training. What is specific training? It is training in the tasks in which the patient needs and wants to be engaged. The first level is daily functional tasks. We should help the patient activate appropriately the muscles needed to accomplish the task, in a goal-directed, task-oriented context. We must consider principles of motor learning as well as currently accepted ideas about how the system controls movements to make our therapy effective. Then, we should take our patients beyond activities of daily living to activities relevant to occupation and recreation.

Because of the current pressures—in the United States at least—for cost containment, it is critical that therapists conduct clinical research studies to demonstrate the effectiveness of therapeutic approaches. We can no longer afford to simply become a disciple of any particular approach, thinking blindly that our way is the best way. We must become responsible to show with controlled studies what is the optimum way, under what conditions, with what diagnoses or symptoms, and in which aged and variously experienced patients. We are currently being forced into interventions to "get the patient out" that are supposedly cost effective. However, we could be training compensatory strategies that will prevent true recovery, thus lengthening the time and increasing the level of care that the person will need in the long run. Are we setting the patient up for complications because we are in so much of a hurry? Yet patients being discharged to home may be beneficial because now therapy can occur in the patient's own environment. The activities may take on more important meaning to the patient, so that he or she will become more engaged and therefore more successful. No one can answer these questions now, but they need to be answered—and soon—so that our patients will be helped in the optimum way, rather than adding to the effects of the original lesion.

REFERENCES

1. Davis WJ. Organizational concepts in the central motor networks of invertebrates. In: Herman RM, Grillner S, Stein PSG, Stuart DG, eds. *Neural Control of Locomotion.* New York: Plenum Press; 1976:265–292.
2. Kelly IF. Reactions of neurons to injury. In: Kandel ER, Schwartz JH, eds. *Principles of Neural Science.* New York: Elsevier/North Holland; 1981:138–146.
3. Nieto-Sampedro M, Cotman CW. Growth factor induction and temporal order in central nervous system repair. In: Cotman CW, ed. *Synaptic Plasticity.* New York: Guilford Press; 1985:407–455.
4. Panayiotou BN, Fotherby MD. Pharmacological therapy for acute stroke: the future. *BJC.* 1995;49(6): 314–317.
5. Shuaib A, Waqaar T, Ijaz MS, et al. Neuroprotection with felbamate: a 7- and 28-day study in transient forebrain ischemia in gerbils. *Brain Res.* 1996;727:65–70.
6. Luria AR, Naydin VL, Tsvetkova LS, et al. Restoration of higher cortical function following local brain damage. In: Vinken RJ, Bruyn GW, eds. *Handbook of Clinical Neurology, Vol. 3.* Amsterdam: North Holland Publishing; 1969:368–433.
7. von Monakow C. Diaschisis. In: Pribram RH, ed. *Brain and Behavior: I. Mood States and Minds.* Baltimore: Penguin; 1969:27–36.
8. Goldberger ME, Murray M. Recovery of function and anatomical plasticity after damage to the adult and neonatal spinal cord. In: Cotman CW, ed. *Synaptic Plasticity.* New York: Guilford Press; 1985:77–110.
9. Le Vere TE. Recovery of function after brain damage: a theory of the behavioral deficit. *Physiol Psychol.* 1980;8:297–308.
10. Taub E. *Constraint induced movement therapy: a new family of treatments for rehabilitation of movement after stroke.* Paper presented at the Teachers College, Columbia University Conference on Movement Disorders Associated with Aging, April 5, 1997, New York.
11. Schallert T, Kozlowski DA, Humm JL, et al. Use-dependent structural events in recovery of function. *Brain Plasticity, Adv Neurol.* 1997;73:229–238.
12. Nudo RJ, Wise BM, SiFuentes F, et al. Neural substrates for the effects of rehabilitative training on motor recovery after ischemic infarct. *Sci.* 1996;272:1791–1794.
13. Bjorklund A, Gage FH, Dunnett SB, et al. Regenerative capacity of central neurons as revealed by intracerebral grafting experiments. In: Bignami A, Bloom FE, Bolis CL, Adeloye A, eds. *Central Nervous System Plasticity and Repair.* New York: Raven Press; 1985:57–62.
14. Dunnett SB, Bjorklund A, Stenevi U. Dopamine-rich transplants in experimental parkinsonism. *Trends Neural Sci.* July 1983:266–270.
15. Kiester E Jr. Spare parts for damaged brains. *Sci 86.* March 1986:33–38.
16. Ahlskog E. Cerebral transplantation for Parkinson's disease: current progress and future prospects. *Mayo Clin Proc.* 1993;68:578–591.
17. Lopez-Lozano J, Braavo G, Brera B, et al. Long-term improvement in patients with severe Parkinson's disease after implantation of fetal ventral mesencephalic tissue in a cavity of the caudate nucleus: 5-year follow up in 10 patients. *J Neurosurg.* 1997;86:931–942.
18. Stein DG, Labbe R, Firl A Jr, et al. Behavioral recovery following implantation of fetal brain tissue into mature rats with bilateral cortical lesions. In: Bjorklund A, Stenevi U, eds. *Neural Grafting in the Mammalian CNS.* New York: Elsevier; 1985:605–614.

19. Dunnett SB, Ryan CN, Levin PD, Reynolds M, et al. Functional consequences of embryonic neocortex transplanted to rats with prefrontal cortex lesions. *Behav Neurosci.* 1987;101:489–503.
20. Bender NK. The stroke therapy research boom: implications for hospital formulary. *Pharmacother.* 1998;18(3 pt 2):108S-115S.
21. Faugier-Grimaud, Frenois, Stein. Cited by Finger S, Stein DG. *Brain Damage and Recovery.* New York: Academic Press; 1982.
22. Levi-Montalcini R, Calissano P. The nerve growth factor. *Sci Am.* 1979;176:297–310.
23. Hart T, Chaimas N, Moore RY, et al. Effects of nerve growth factor on behavioral recovery following caudate nucleus lesions in rats. *Brain Res Bull.* 1978;3:245–250.
24. Feeney DM, Sutton RL. Pharmacotherapy for recovery of function after brain injury. *Crit Rev Neurobiol.* 1987;3:135–197.
25. Karpiak SE. Exogenous gangliosides enhance recovery from CNS injury. In: Ledeen RW, Yu RK, Rapport MM, Suzuki K, eds. *Ganglioside Structure, Function and Biomedical Potential.* New York: Plenum Press; 1984:489–497.
26. Sabel BA, Slavin MD, Stein DG. GM1 ganglioside treatment facilitates behavioral recovery from bilateral brain damage. *Sci.* 1984;225:340–342.
27. Riggott M, Matthew W. Neurite outgrowth is enhanced by anti-idiotypic monoclonal antibodies to the ganglioside GM1. *Exp Neurol.* 1997;145:278–287.
28. Ward AA Jr, Kennard MA. Effects of cholinergic drugs on recovery of function following lesions of the central nervous system. *Yale J Biol Med.* 1942;15:189–228.
29. Meyer FM, Horel JA, Meyer DR. Effects of d-amphetamine upon placing responses in neodecorticate cats. *J Comp Physiol Psychol.* 1963;56:402–404.
30. Meyer DR, Meyer PM. Dynamics and bases of recoveries of functions after injuries to the cerebral cortex. *Physiol Psychol.* 1977;5:133–165.
31. Crisostomo EA, Duncan PW, Propst MA, et al. Evidence that amphetamine with physical therapy promotes recovery of motor function in stroke patients. *Ann Neurol.* 1988;23:94–97.
32. Goldman PS. An alternative to developmental plasticity: heterology of CNS structures in infants and adults. In: Stein DG, Rosen JJ, Butters N, eds. *Plasticity and Recovery of Function in the Central Nervous System.* New York: Academic Press; 1974:149–174.
33. Milner B. Sparing of language functions after early unilateral brain damage. *Neurosci Res Prog Bull.* 1974;12:213–217.
34. Hicks SF, D'Amato CJ. Motor-sensory and visual behavior after hemispherectomy in newborn and mature rats. *Exp Neurol.* 1970;29:416–438.
35. Kennard MA. Relation of age to motor impairment in man and in sub-human primates. *Arch Neurol Psychiatry.* 1940;44:377–397.
36. Levin HS, Ewing-Cobbs L, Benton AL. Age and recovery from brain damage: a review of clinical studies. In: Scheff SW, ed. *Aging and Recovery of Function in the Central Nervous System.* New York: Plenum Press; 1984.
37. Stein DC, Firl A. Brain damage and reorganization of function in old age. *Exp Neurol.* 1976;52:157–167.
38. Caulder SL, Gentile AM. Sex differences in recovery of locomotion following cortical damage in rats. *Soc Neurosci Abstracts.* 1984;10:320.
39. Roof R, Duvdevani R, Stein D. Gender influences outcome of brain injury: progesterone plays a protective role. *Brain Res.* 1993;607:333–336.
40. Finger S. Lesion momentum and behavior. In: Finger S, ed. *Recovery from Brain Damage.* New York: Plenum Press; 1978:135–164.
41. Finger S, Walbran B, Stein DC. Brain damage and behavioral recovery: serial lesion phenomena. *Brain Res.* 1973;63:1–18.
42. Travis AM, Woolsey CN. Motor performance in monkeys after bilateral partial and total cerebral decortications. *Am J Phys Med.* 1956;35:273–310.
43. Hebb D. *The Organization of Behavior.* New York: John Wiley & Sons; 1949:298–299.

44. Weisel TN, Hubel DH. Comparison of the effects of unilateral and bilateral eye closure on cortical unit responses in kittens. *J Neurophysiol.* 1965;28:1029–1040.
45. Rosenzweig MR, Bennett EL. Effects of differential environments on brain weights and enzyme activities in gerbils, rats and mice. *Dev Psychol.* 1969;2:87–95.
46. Smith CJ. Mass action and early environment. *J Comp Physiol Psychol.* 1959;52:154–156.
47. Hughes KR. Dorsal and ventral hippocampus lesions and maze learning: influence of preoperative environment. *Can J Psychol.* 1965;19:325–332.
48. Donovick PJ, Burright RC, Swidler MA. Presurgical rearing environment alters exploration, fluid consumption, and learning of septal lesioned and control rats. *Physiol Behav.* 1973;11:543–553.
49. Held JM, Gordon J, Gentile AM. Environmental influences on locomotor recovery following cortical lesions in rats. *Behav Neurosci.* 1985;99:678–690.
50. Schwartz S. Effects of neonatal cortical lesions and early environmental factors on adult rat behavior. *J Comp Physiol Psychol.* 1964;57:72–77.
51. Will BE, Rosenzweig MR, Bennet EL, et al. Relatively brief environmental enrichment aids recovery of learning capacity and alters brain measures after postweaning brain lesions in rats. *J Comp Physiol Psychol.* 1977;91:33–50.
52. Einon DF, Morgan MF, Will BE. Effects of post-operative environment on recovery from dorsal hippocampal lesions in young rats: test of spatial memory and motor transfer. *Q J Exp Psychol.* 1980;32:137–148.
53. Finger S. Environmental attenuation of brain lesion symptoms. In: Finger S, ed. *Recovery from Brain Damage.* New York: Plenum Press; 1978:297–330.
54. Kalra L. The influence of stroke unit rehabilitation on functional recovery from stroke. *Stroke.* 1994;25:821–825.
55. Weese CD, Neimand D, Finger S. Cortical lesions and somesthesis in rats: effects of training and overtraining prior to surgery. *Exp Brain Res.* 1973;16:542–550.
56. Goldberger ME. Recovery of movement after CNS lesions in monkeys. In: Stein DG, Rosen JJ, Butters N, eds. *Plasticity and Recovery of Function in the Central Nervous System.* New York: Academic Press; 1974:265–338.
57. Ogden R, Franz SI. On cerebral motor control: the recovery from experimentally produced hemiplegia. *Psychobiology.* 1917;1:33–49.
58. Black P, Markowitz RS, Cianci SN. Recovery of motor function after lesions in motor cortex of monkeys. *Ciba Found Symp.* 1975;34:65–83.
59. Gentile AM, Green S, Nieburg A, et al. Disruption and recovery of locomotor and manipulatory behavior following cortical lesions in rats. *Behav Biol.* 1978;22:417–455.
60. Isaacson RL, Spear LP. A new perspective for the interpretation of early brain damage. In: Finger S, Almli CR, eds. *Early Brain Damage.* New York: Academic Press; 1984;2:73–98.
61. Lashley RS. Studies of cerebral function in learning: V. The retention of motor habits after destruction of the so-called motor areas in primates. *Arch Neurol Psychiatry.* 1924;12:249–276.

Index